Holt Mathematics

Chapter 8 Resource Book

HOLT, RINEHART AND WINSTON
A Harcourt Education Company
Orlando • Austin • New York • San Diego • London

Printed in the United States of America

ISBN 0-03-078224-4

6 7 170 09 08

CONTENTS

Blackline Masters

Date __________

Dear Family,

In this chapter, your child will learn some of the fundamentals of plane geometry. This includes learning how to describe various figures in geometric terms, measure angles, and solve problems involving measurements of angles and lines. It also involves learning about polygon relationships. The information in this chapter can be applied to careers in aviation, navigation, design, and many others.

The building blocks of geometry are **points, lines,** and **planes.**

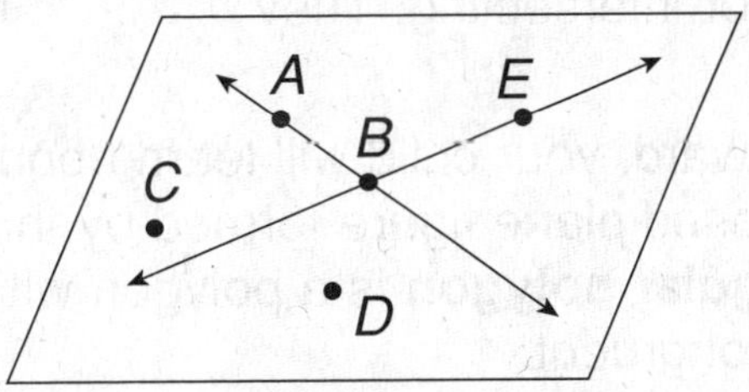

Use the figure to name three points.
A, *C*, and *D*
Five points are labeled:
points *A*, *B*, *C*, *D*, and *E*.

Use the figure to name two lines.
$\overleftrightarrow{AB}$ and $\overleftrightarrow{BE}$
You can also write $\overleftrightarrow{BA}$ and $\overleftrightarrow{EB}$.

Use the figure to name a point shared by two lines.
Point *B* is a point on $\overleftrightarrow{AB}$ and $\overleftrightarrow{BE}$.

Your child will learn to measure an **angle** using a protractor. Two rays with a common endpoint form an angle. Different angle relationships that will help us solve everyday problems will be investigated.

Find the unknown angle measure.
The angles are complementary.

$$\begin{aligned} 55^\circ + A &= 90^\circ \\ -55^\circ \quad &\quad -55^\circ \\ A &= 35^\circ \end{aligned}$$

The sum of the measures is 90°.

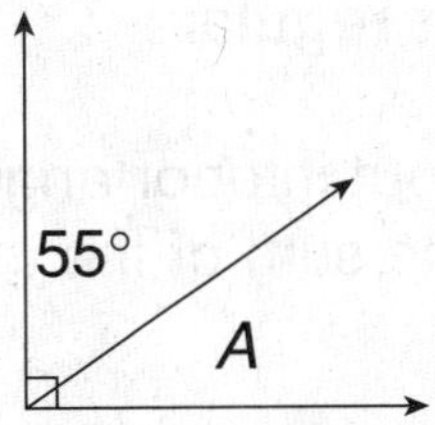

$m\angle A = 35^\circ$

Your child will learn to classify the different types of **lines.**

Classify each pair of lines.

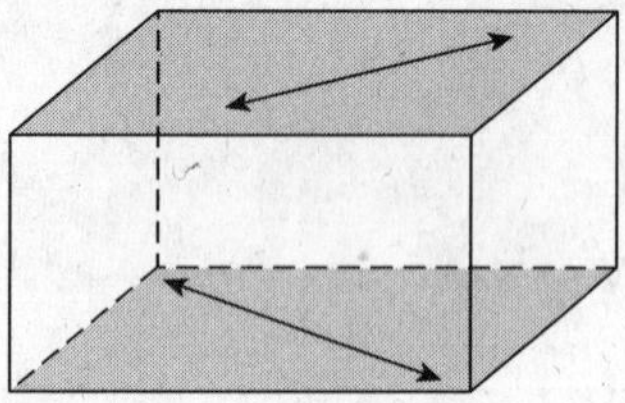

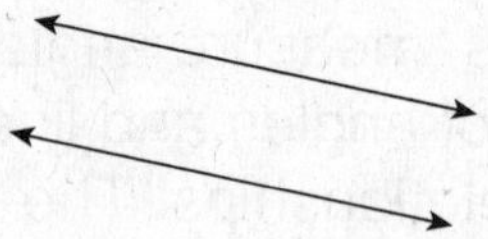

The lines are in different planes and are not parallel or intersecting. They are skew.

The lines are in the same plane. They do not appear to intersect. They are parallel.

As we move forward, your child will learn about polygons. A **polygon** is a closed plane figure formed by three or more line segments. A **regular polygon** is a polygon with all sides congruent and all angles congruent.

Name each polygon and tell whether it appears to be *regular* or *not regular.*

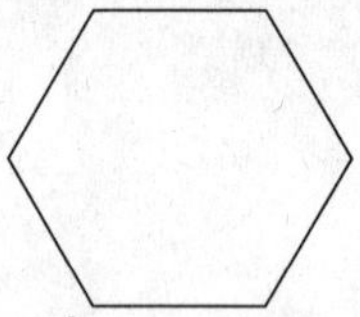

There are 6 sides and 6 angles.
hexagon
The sides and angles appear to be congruent.
regular

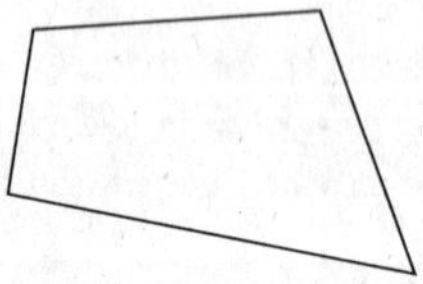

There are 4 sides and 4 angles.
quadrilateral
The sides do not appear to be congruent.
not regular

Your child will learn about interior angles. The sum of the angles in a triangle is 180°, so the sum of the interior angles in a quadrilateral is 360°.

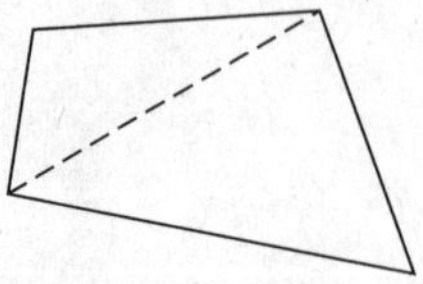

For additional resources, visit go.hrw.com and enter the keyword MR7 Parent.

Name ______________________ Date ____________ Class ____________

LESSON 8-1

Practice A

Building Blocks of Geometry

Name each geometric figure.

1. •*P*

____

2. 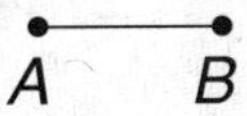

____

3. 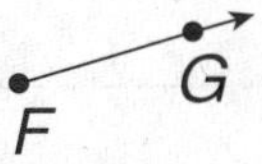

____

4. 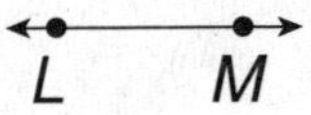

____

5. 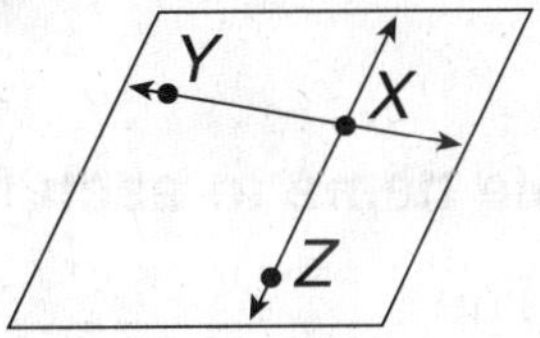

____

6. 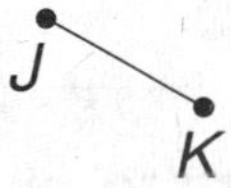

____

Use the diagram to choose the correct answer.

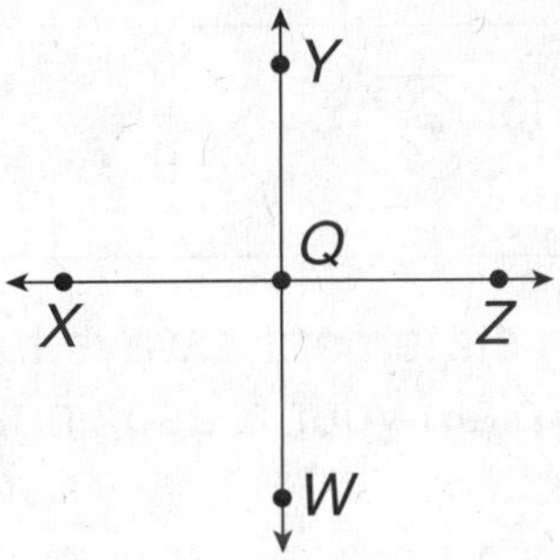

7. Which of the following does not name a line on the diagram?

A Line *YQ*

B Line *WZ*

C $\overleftrightarrow{ZX}$

D $\overleftrightarrow{QW}$

8. Which two points cannot be used to name a ray on the diagram?

F *Q* and *Y*

G *Z* and *X*

H *X* and *Y*

J *W* and *Q*

9. How many points are needed to define a line or line segment?

____

10. Lynn named this ray $\overrightarrow{PQ}$. Explain her error and give the correct name for the ray.

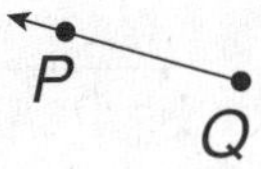

____

____

Name ______________________ Date ____________ Class ____________

LESSON 8-1

Practice B

Building Blocks of Geometry

Use the diagram to name each geometric figure.

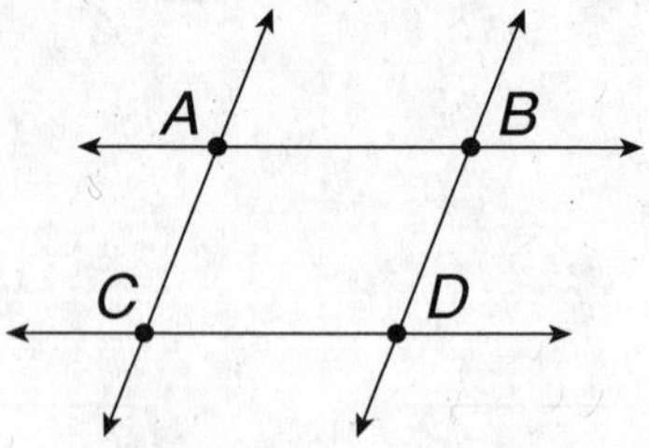

1. two points ____________

2. a plane ____________

3. a line segment ____________

4. a point shared by two lines ____________

5. a line ____________

Use the diagram to give a possible name to each figure.

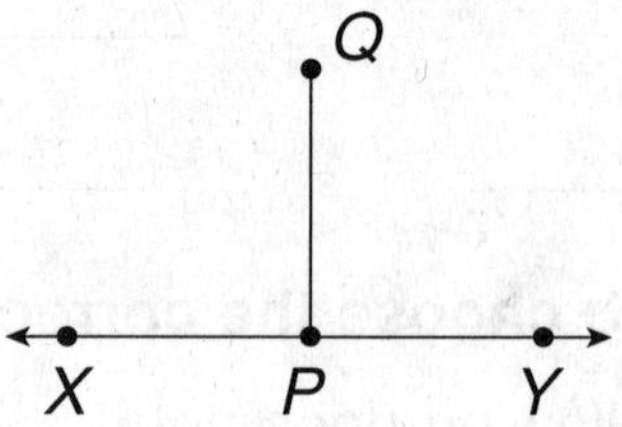

6. two different ways to name the line

____

7. four different names for rays

____

8. another name for $\overline{QP}$

____

9. Is the following statement always true, sometimes true, or never true? Explain your reasoning. A line is longer than a line segment.

____

____

____

10. Using endpoints as your basis, explain how a line, a line segment, and a ray are different.

____

____

____

Name ______________________ Date ____________ Class ____________

LESSON 8-1

Practice C

Building Blocks of Geometry

Use the diagram to name each geometric figure.

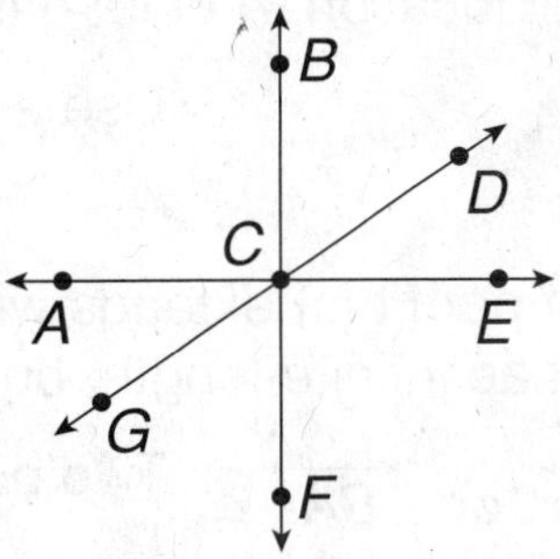

1. a point shared by three lines ______

2. three points on the same line

____

3. four different rays

____

4. four different names for the same line segment

____

____

Draw and label an example of each geometric figure.

5. $\overrightarrow{PQ}$

6. $\overleftrightarrow{ST}$

7. $\overline{XY}$

8. plane *FGH*

9. If 3 points are labeled on a line, how many different line segments can be named on that line? How many different rays?

____

10. Name every geometric figure that is part of line *AB*.

____

Name ______________________ Date ____________ Class ____________

LESSON **8-1**

Reteach

Building Blocks of Geometry

Here are terms that can help you understand geometry.

A point is an exact location. A point has no size.

• *A* point *A* or *A* — Use a capital letter to name a point.

A line is a straight path that extends without end in opposite directions. A line has infinite length, but no width.

$\overleftrightarrow{AB}$ or $\overleftrightarrow{BA}$ — Two points name a line.

A plane is a flat surface that extends without end in all directions. A plane has infinite length and width, but no depth.

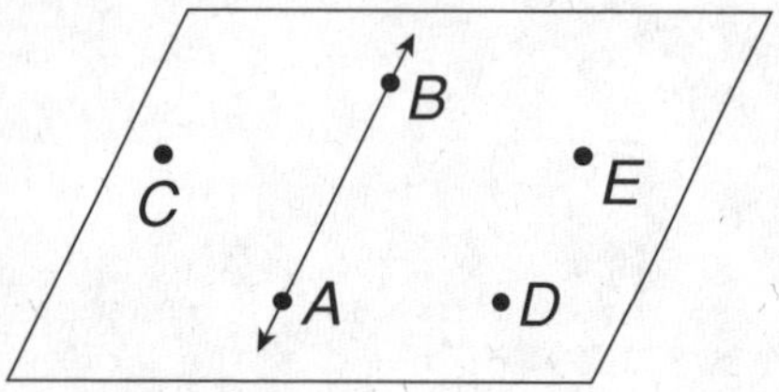

CDE *EDC*
DEC *CED*
ECD *DCE*

A plane is named by 3 points on the plane that are not on a line.

A line segment has two endpoints. The length of a line segment can be measured.

$\overline{JK}$ or $\overline{KJ}$ — A line segment is named by its endpoints.

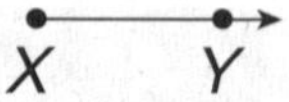

$\overrightarrow{XY}$

A ray has one endpoint and extends without end in one direction. A ray is named by its endpoint first and another point on the ray.

Use the diagram to name each geometric figure.

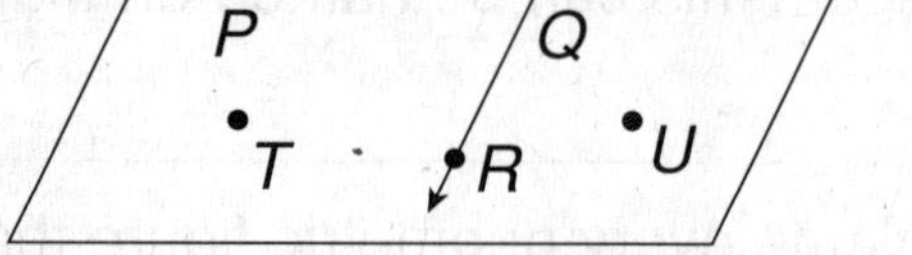

1. two points

____

2. two lines

____

3. a plane

____

4. two line segments

____

5. two rays

____

6. a point shared by two lines

____

Name ______________________ Date ____________ Class ____________

LESSON **8-1**
Challenge
Points of Light

Astronomers have divided the sky into 88 constellations, or groups of stars. To map these constellations, astronomers use a point for each star and draw imaginary lines between the points. One of the easiest groups of stars to see is called the Big Dipper. The Big Dipper is not officially a constellation. It is only part of a large constellation called Ursa Major, or Great Bear.

Imagine the stars of Ursa Major are labeled as shown below. Follow these instructions to make a star map of Ursa Major.

1. Use line segments to connect points *A–G* in alphabetical order, and then draw $\overline{DG}$.
2. Use line segments to connect points *G–L* in alphabetical order, and then draw $\overline{FJ}$.
3. Use line segments to connect points *M–O* in alphabetical order. Then draw $\overline{EM}$ and $\overline{MP}$.
4. Label the star grouping formed by the 7 points *A–G* as the "Big Dipper."
5. Then label each point of the Big Dipper with its proper star name: *A*: Alkaid; *B*: Mizar; *C*: Alioth; *D*: Megrez; *E*: Phecda; *F*: Merak; *G*: Dubhe

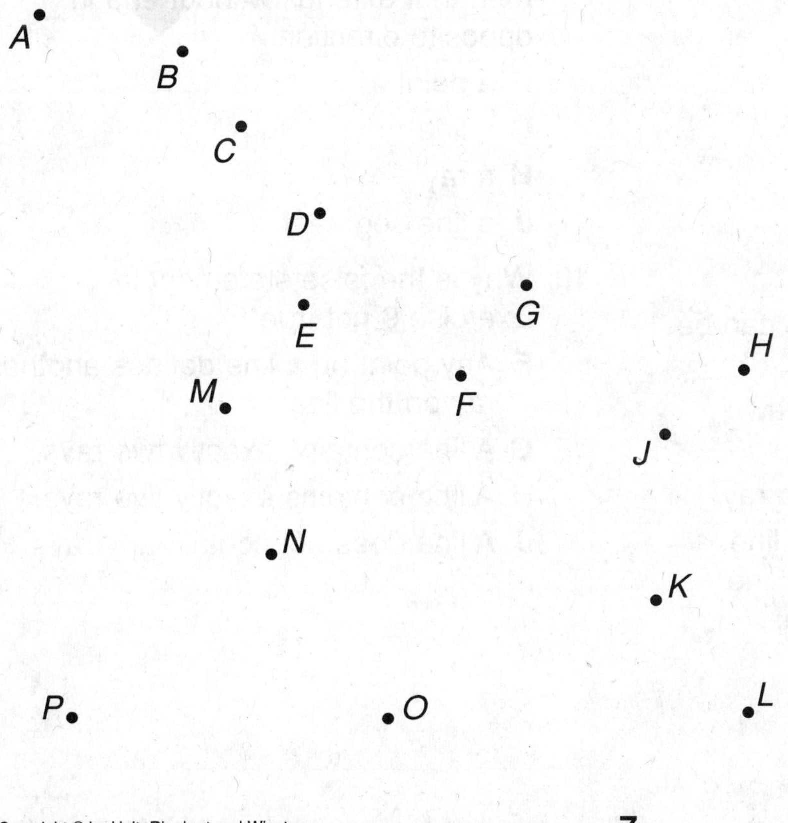

Name ______________________ Date ___________ Class ___________

LESSON **8-1**

Problem Solving

Building Blocks of Geometry

Place your hand down flat on a sheet of paper. Draw a point at the tip of your thumb, the tip of your middle finger, and the tip of your pinky.

1. Label the thumb point *A*, the middle finger point *B*, and the pinky point *C*.

____

2. Name all the planes you possibly can with points *A*, *B*, and *C*.

____

3. Draw and name all the lines you can make with points *A*, *B*, and *C*.

____

4. Name all the line segments possible using points *A*, *B*, and *C*.

____

5. Name all the rays possible using points *A*, *B*, and *C*.

____

____

6. Choose one line that you drew. Give all the different possible names for that line.

____

____

Circle the letter of the correct answer.

7. Which of the following has exactly one endpoint?

A $\overleftrightarrow{OP}$

B $\overline{AB}$

C $\overleftrightarrow{TR}$

D $\overrightarrow{SM}$

8. Which of the following is a straight path that extends without end in opposite directions?

F a point

G a line

H a ray

J a line segment

9. Which statement is false?

A An infinite number of lines can be drawn through one point.

B Exactly one line can be drawn between two points.

C A line contains exactly one ray.

D If points *A* and *B* are on a line, then line segment *AB* and line segment *BA* are the same.

10. Why is the false statement in Exercise 9 not true?

F Any point on a line defines another ray on the line.

G A line contains exactly two rays.

H A line contains exactly five rays.

J A line does not contain any rays.

Name ______________________ Date ____________ Class ____________

LESSON 8-1

Reading Strategies

Understanding Symbols

A **point** marks an exact location in space. Read: "point *T*."

A capital letter is used to name a point.

A **line** is a straight path that extends forever in both directions. Read: "line *DR*" or "line *RD*."

Two points on the line name a line.

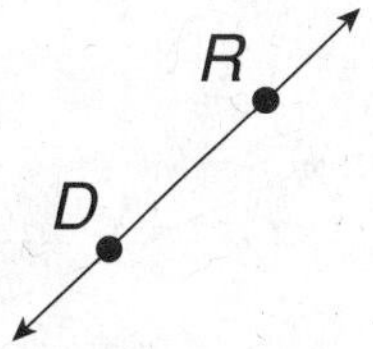

A **ray** is part of a line that has one endpoint and goes on forever in one direction. Read: "ray *SM*."

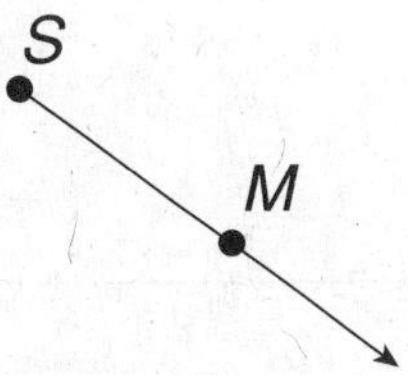

A **line segment** is part of a line between two endpoints.

Use the two endpoints to name the line.
Read: "segment *VW*" or "segment *WV*."

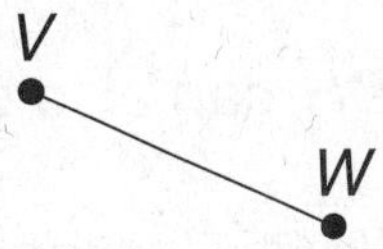

Use this figure to answer the questions.

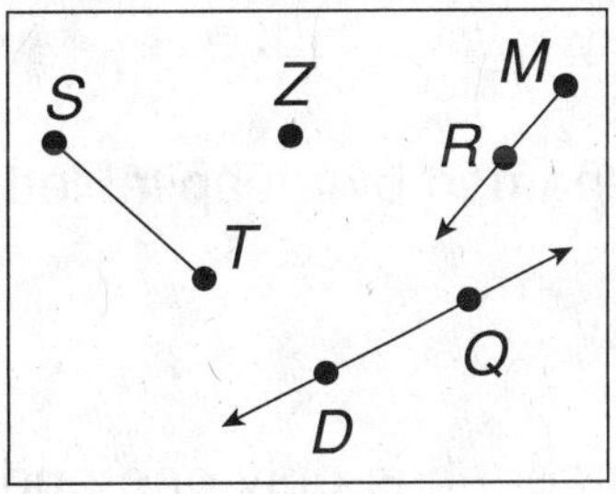

1. What symbols would you use to name the line segment? ____________

2. What symbols would you use to name the line? ____________

3. What symbols would you use to name the ray? ____________

4. Write the name of the point. ____________

5. What is the difference between a line and a ray?

____

____

Name ______________________ Date ____________ Class ____________

LESSON 8-1

Puzzles, Twisters & Teasers

Is it the Truth?

Determine which statements are true. Rearrange the letters of the true statements to solve the riddle.

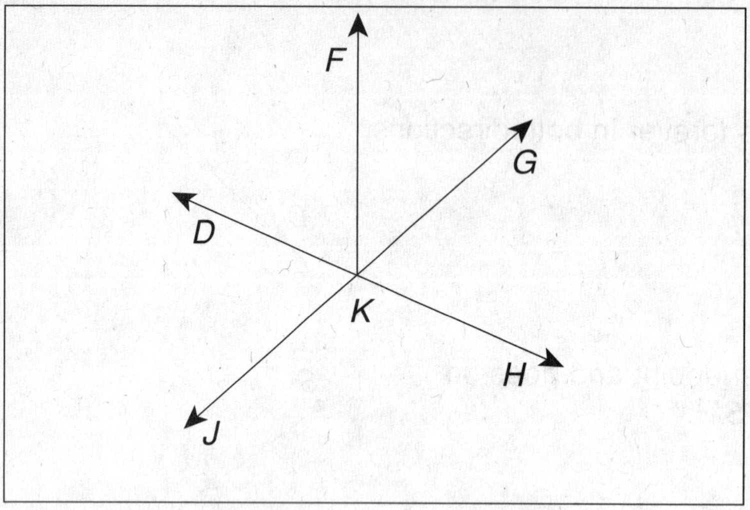

1. R Point *G* is on $\overleftrightarrow{JG}$.
2. O $\overleftrightarrow{DH}$ can also be written $\overleftrightarrow{HD}$.
3. B *JKG* is a plane.
4. S *GK* is a ray.
5. A A point is named by a capital letter.
6. N *KD* is a ray.
7. U A line segment has only one end point.
8. T *FKG* is a plane.
9. C A line is a straight path that extends without end in all directions.
10. O A ray is named by its endpoint first.

What do you call a song played on an automobile radio?

A ___ ___ ___ ___ ___ ___ ___

Name ______________________ Date __________ Class __________

LESSON 8-2

Practice A

Measuring and Classifying Angles

Name the type of angle shown.

1.

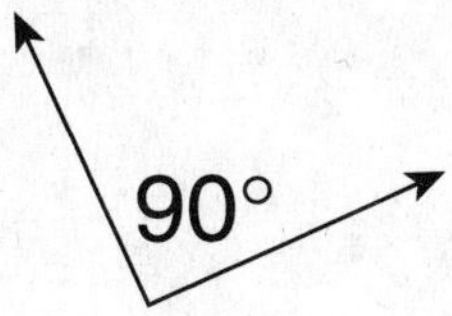

____

2.

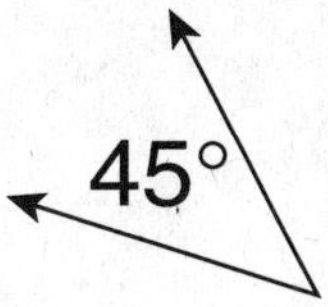

____

3.

____

4.

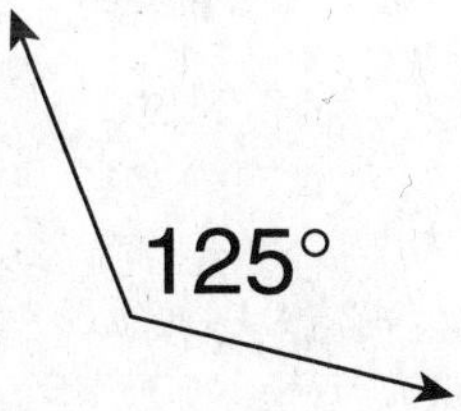

____

5.

____

6.

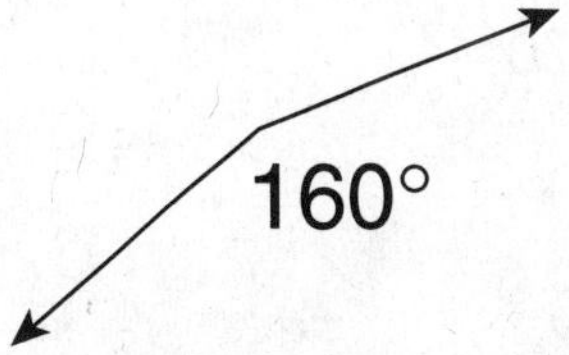

____

Use a protractor to measure each angle.

7.

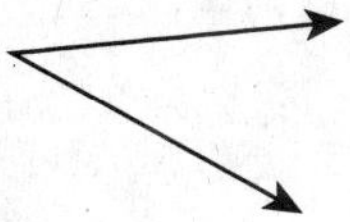

____

8.

____

9.

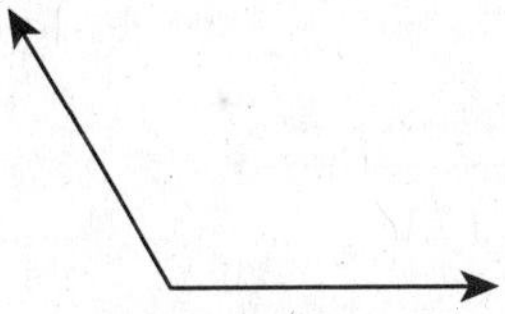

____

10. Carl is making a desk for his office. One of the angles of the desk measures greater than 90°. What two types of angles could that angle be?

____

11. How many angles make up the capital letter L? What types of angles are they?

____

Name ______________________ Date ____________ Class ____________

LESSON **8-2**

Practice B
Measuring and Classifying Angles

Use a protractor to measure each angle.

1.

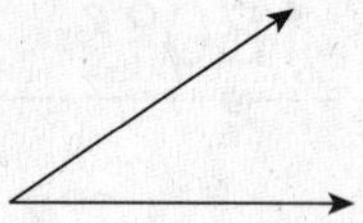

2.

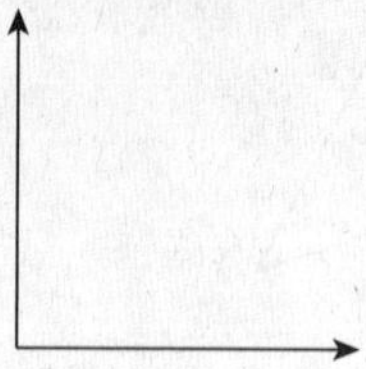

3.

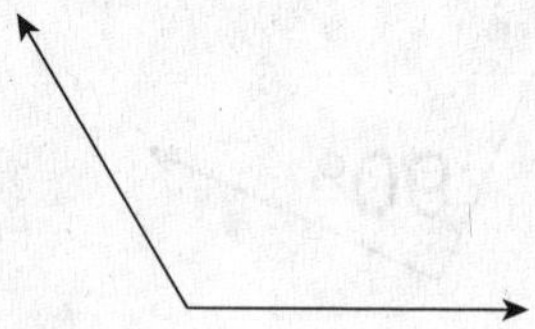

______________ ______________ ______________

Use a protractor to draw an angle with each given measure.

4. 70°

5. 115°

6. 45°

Classify each angle as acute, right, obtuse, or straight.

7.

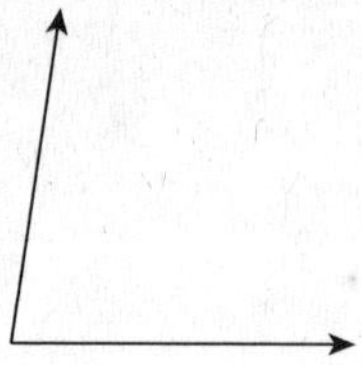

8.

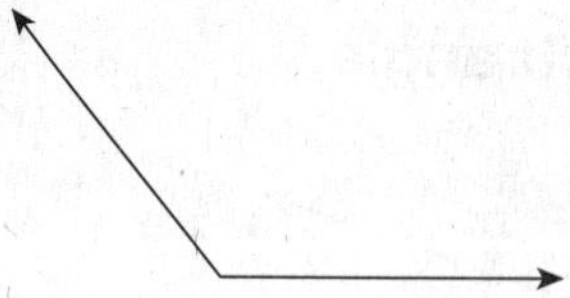

9.

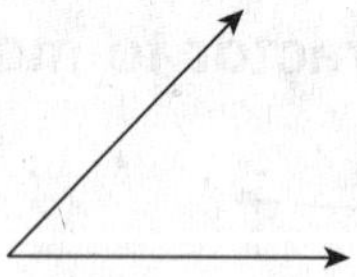

______________ ______________ ______________

10. The frame for a kite has two angles that together form a right angle. What type of angle is each of those angles? Explain.

____

____

11. What kinds of angles are in each of the letters in this word? **TAXI**

____

____

Name ______________________ Date __________ Class __________

LESSON **8-2**

Practice C
Measuring and Classifying Angles

Use a protractor to measure each angle and name the type of angle.

1.
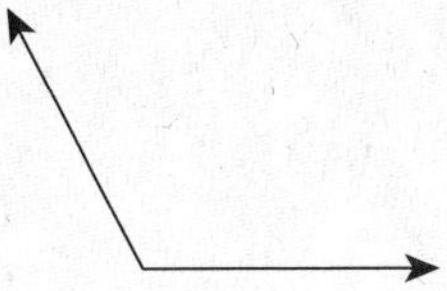

2.
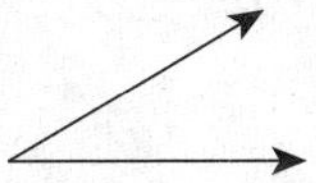

3.
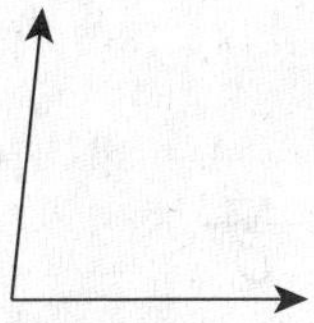

Use a protractor to draw each angle.

4. an acute angle whose measure is greater than 45°

5. an obtuse angle whose measure is between 95° and 120°

6. an obtuse angle formed by a right angle and an acute angle

Classify each angle as acute, right, obtuse, or straight.

7.
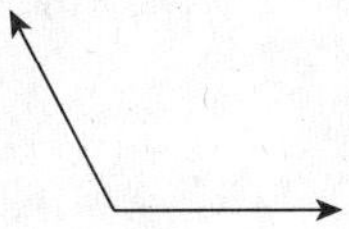

8.
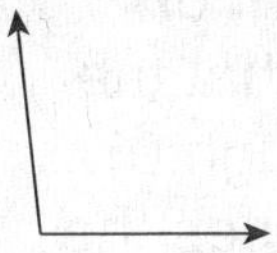

9.
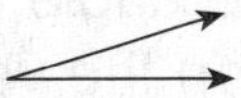

10. Together, ∠1 and ∠2 form a straight angle. If ∠1 is acute, what must be true about ∠2? Explain.

11. Is the following statement always true, sometimes true, or never true? Explain your reasoning. Two acute angles of the same size form a right angle.

Name ______________________ Date ___________ Class ___________

LESSON 8-2

Reteach

Measuring and Classifying Angles

An angle is formed by two rays with a common endpoint, called the vertex.

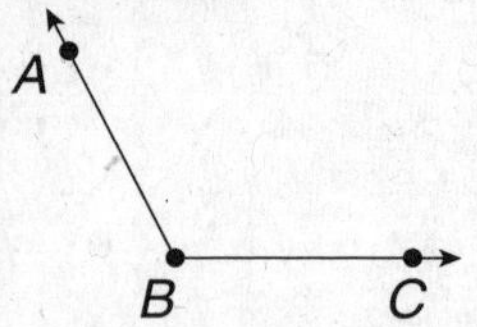

You can name an angle in different ways.

By its vertex	OR	By its vertex and one point from each ray.
$\angle B$		$\angle ABC$, $\angle CBA$ Vertex letter in the middle.

Give three ways to name each angle.

1.

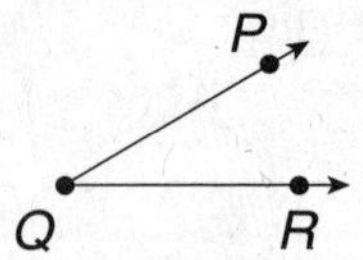

____

2.

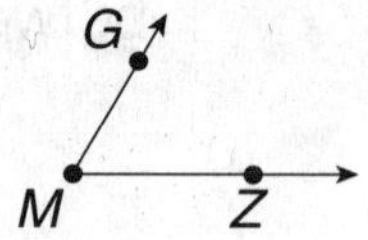

____

You can use a protractor to measure an angle.

To measure the angle, place the center point of the protractor on the vertex, *K*. Make sure that the protractor is placed so that $\overrightarrow{KL}$ passes through 0°. The measure of the angle is where $\overrightarrow{KJ}$ crosses the protractor.

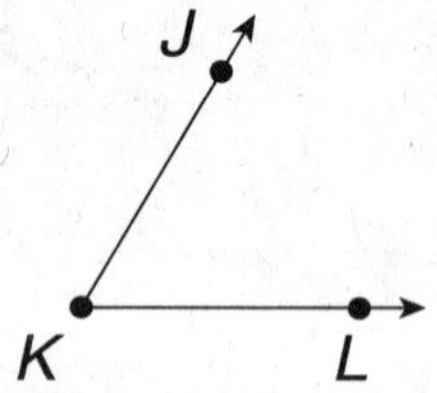

The measure of the angle is 60 degrees.

Write $m\angle K = 60°$.

Use a protractor to measure each angle.

3.

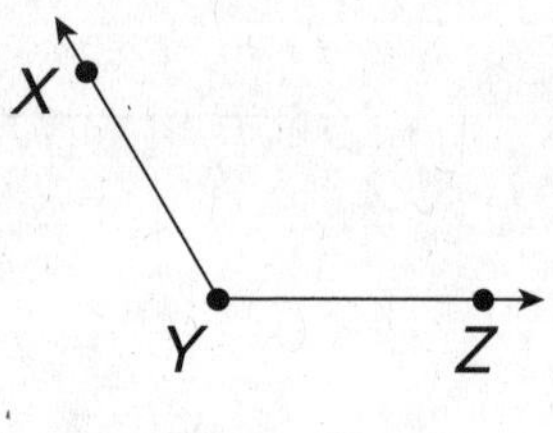

____

4.

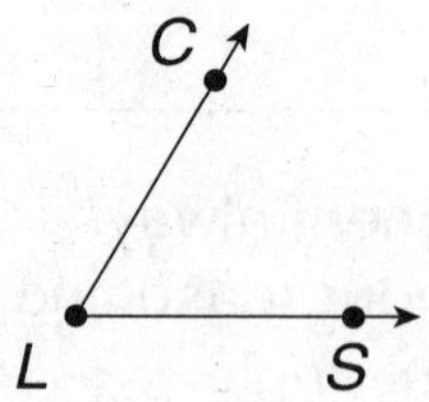

____

Name ______________________ Date __________ Class __________

LESSON **8-2**

Reteach

Measuring and Classifying Angles (continued)

You can use a protractor to draw an angle.

To draw ∠*PQR* that measures 145°, first draw a base ray. Label the endpoint *Q*. Next, arrange the protractor so that its center is on *Q* and the base ray passes through 0°. Then, make a mark at 145° above the scale on the protractor. Use a straightedge to draw a ray from *Q* through the mark you made.

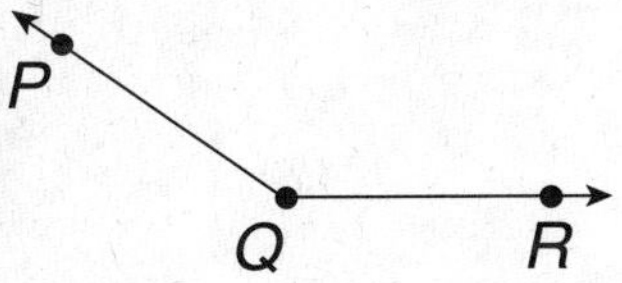

Label one point on a ray *P* and a point on the other ray *R*.

Use a protractor to draw an angle with each given measure.

5. 50°

6. 170°

7. 85°

8. 110°

You can classify an angle by its measure.

If the measure is:	< 90°	> 90°	= 90°	= 180°
Then the angle is:	acute	obtuse	right	straight

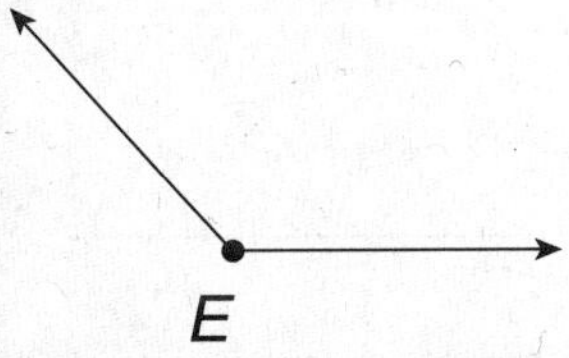

This angle is an obtuse angle because its measure is greater than 90°

Classify each angle.

9.

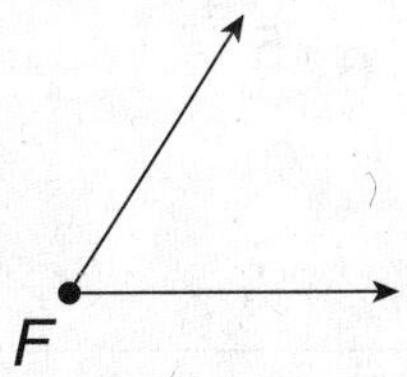

____

10.

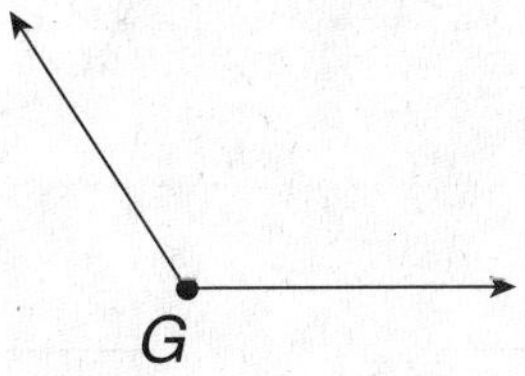

____

Name ______________________ Date ____________ Class ____________

LESSON **8-2**

Challenge

I'll Meet You in 30°

No one knows for sure when people invented clocks to measure time. The first clocks were probably sundials, which measure time by the sun's shadow moving around a circle. Although today's clocks no longer use the sun to measure time, many people still use a circle to display time. A circle has 360°, and a clock face has 12 hours marked. So, each hour on the clock represents 30° of the circle.

Write the time shown on each clock below. Then use a protractor to measure each angle formed by the clock's hands and tell what kind of angle it is.

1.

____

____

2.

____

____

3.

____

____

4.

____

____

5.

____

____

6.

____

____

7.

____

____

8.

____

____

9.

____

____

Name ______________________ Date ____________ Class ____________

LESSON 8-2

Problem Solving

Measuring and Classifying Angles

Write the correct answer.

1. When a patient is lying flat in a hospital bed, what type of angle does the patient's body form? What is the measurement of that angle?

____

2. When a patient is sitting straight up in a hospital bed, the upper body has been raised to what angle? What type of angle is that?

____

3. Most hospital beds have a setting for the Fowler position. In this position, the patient's upper body is raised to form a 60° to 70° angle from a flat position. What types of angles are these?

____

4. What are the greatest and least differences between the straight-up position and the Fowler position in a hospital bed?

____

Circle the letter of the correct answer.

5. Medical technicians often set the handles of crutches so that the patient's elbow is at a 30° angle. What type of angle is this?

A acute angle
B right angle
C obtuse angle
D straight angle

6. By law, wheelchair ramps in public places cannot be greater than 5 degrees. Which type of angle does a wheelchair ramp in public form with the ground?

F acute angle
G right angle
H obtuse angle
J straight angle

7. Physical therapists use a goniometer to measure the extension of a sitting patient's knee. Resting is 90°, and full extension is 180°. What angle does the goniometer measure if the patient's knee is at $\frac{1}{2}$ extension?

A 45°
B 90°
C 135°
D 0°

8. The Q-angle is measured between two points on a patient's hip joint and one point on the knee joint. A normal Q-measure for men is 14° plus or minus 3 degrees. What type of angle is any normal Q-angle for men?

F straight
G obtuse
H right
J acute

LESSON 8-2

Reading Strategies
Use Graphic Aids

Two rays with the same endpoint form an **angle.** The endpoint is called the **vertex.** Three letters can name an angle—a point on each side and the vertex as the middle letter. A single letter, the vertex, can also name it.

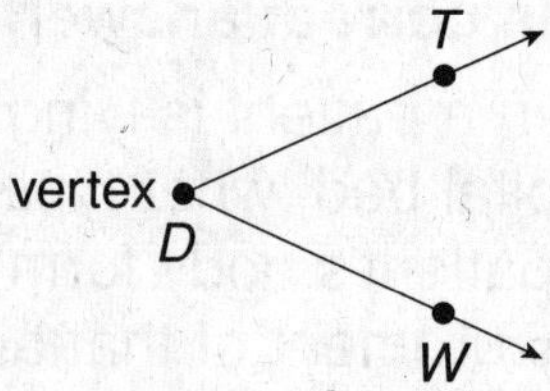

Read: "angle *TDW*," "angle *WDT*," or "angle *D*."

The size or measure of an angle can also name it.

A **right angle** measures exactly 90°. The symbol for a right angle is ⅂.

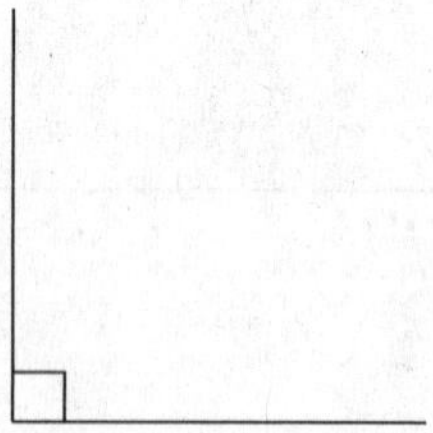

An **acute angle** measures less than 90°.

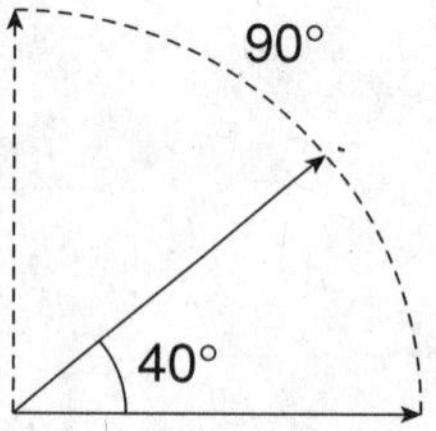

An **obtuse angle** measures more than 90°.

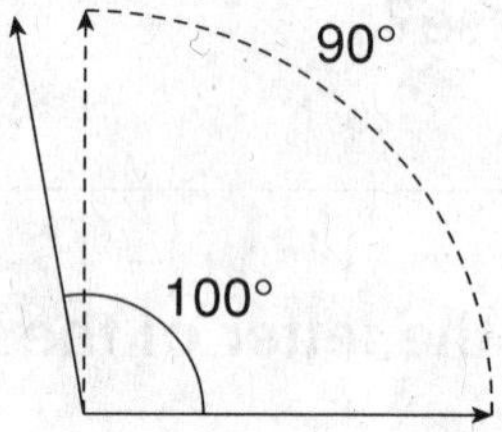

Write *right, acute,* or *obtuse* to name each angle.

1.

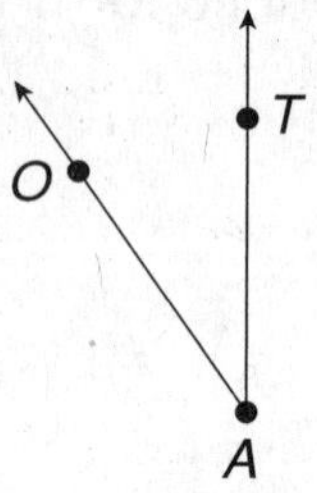

____

2.

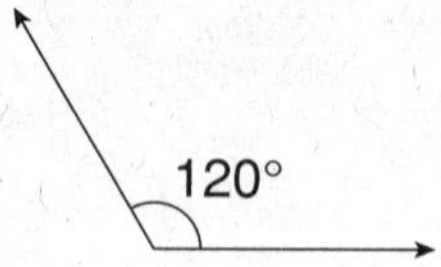

____

3.

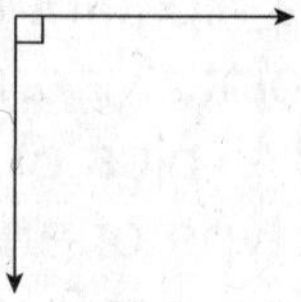

____

4.

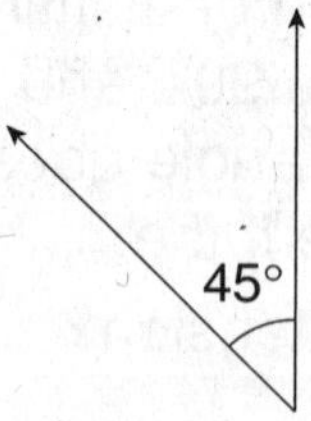

____

Name ______________________ Date ____________ Class ____________

LESSON **8-2**

Puzzles, Twisters & Teasers

What's my Angle?

Measure each angle. Write the letter naming the angle on the line above the correct answer at the bottom of the page. Some letters will not be used.

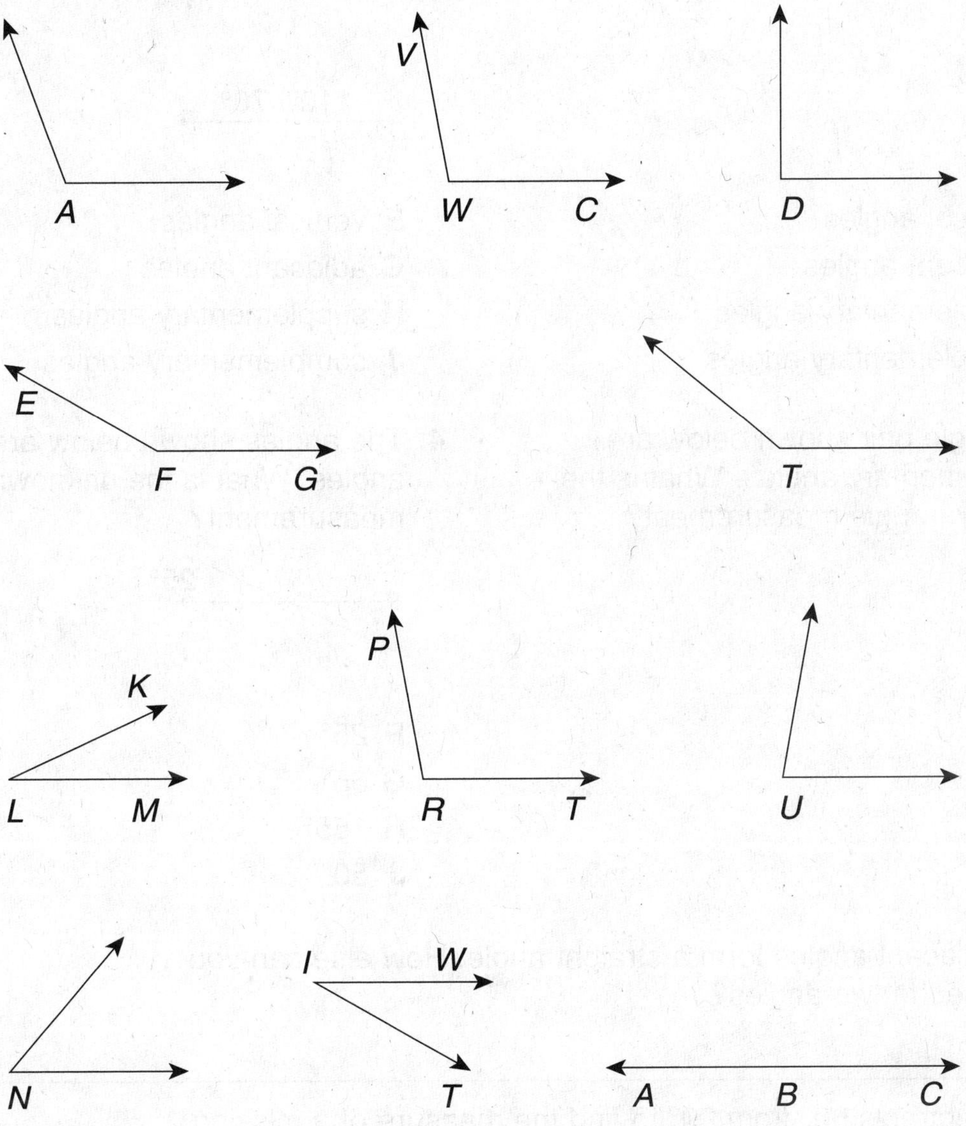

What do you call a boring fish?

____	____	____	____	____	–	____	____	____
110°	90°	80°	25°	25°		150°	30°	50°

Name ______________________ Date ____________ Class ____________

LESSON 8-3

Practice A
Angle Relationships

Choose the correct answer.

1. Which best describes the type of angle pair shown below?

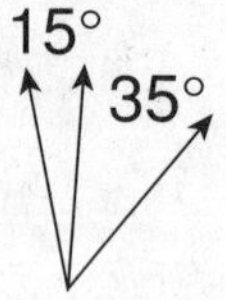

A vertical angles
B adjacent angles
C supplementary angles
D complementary angles

2. Which best describes the type of angle pair shown below?

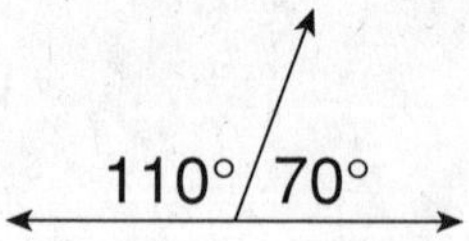

F vertical angles
G adjacent angles
H supplementary angles
J complementary angles

3. The angle pair shown below are complementary angles. What is the unknown angle measurement?

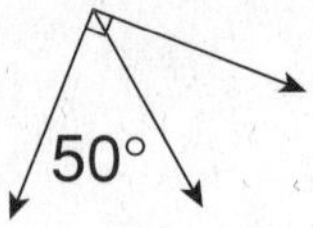

A 50°
B 40°
C 130°
D 90°

4. The angles shown below are vertical angles. What is the unknown angle measurement?

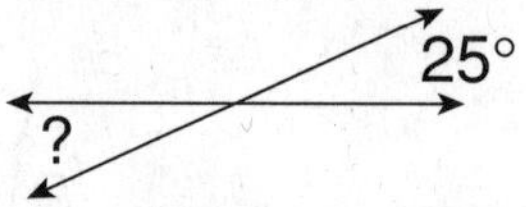

F 25°
G 65°
H 155°
J 50°

5. Two adjacent angles form a straight angle. How else can you describe the two angles?

____

6. Kelly subtracts 60° from 90° to find the measure of a missing angle in an angle pair. What type of angle pair is she working with?

____

Name ______________________ Date ____________ Class ____________

LESSON 8-3

Practice B
Angle Relationships

Identify the type of each angle pair shown.

1.

____

2.

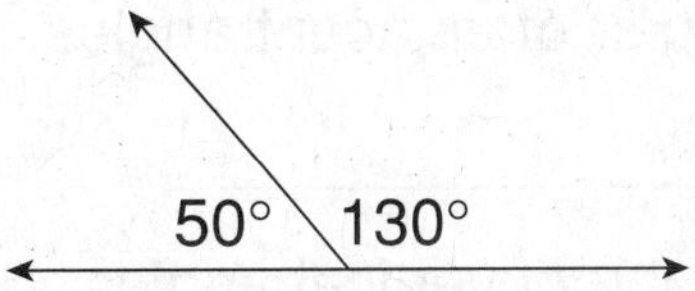

____

3.

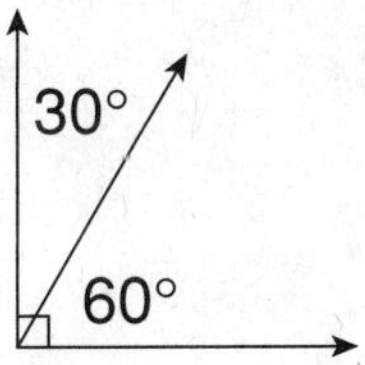

____

4.

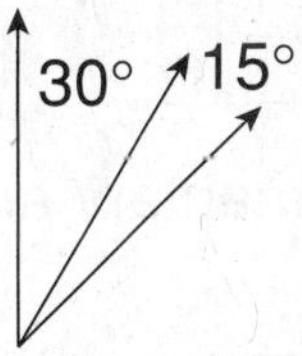

____

Find each unknown angle measure.

5. The angles are supplementary.

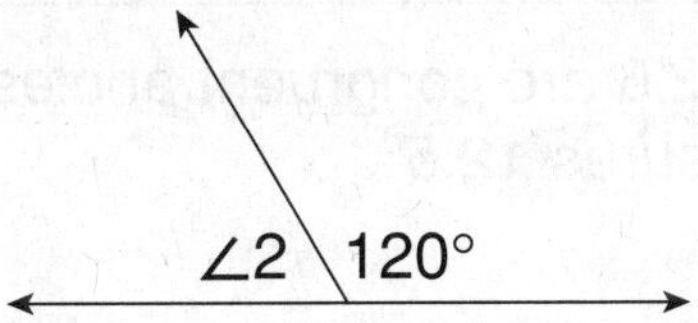

____

6. The angles are complementary.

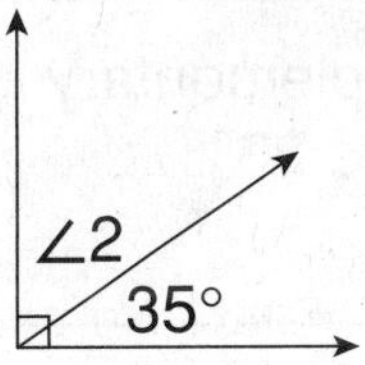

____

7. Anita says the plus sign + forms 2 pairs of vertical angles. Charles says it forms 2 pairs of congruent angles. Who is correct? Explain.

____

____

____

8. Is the following statement always true, sometimes true, or never true? Explain your reasoning. Two congruent angles that are complementary both measure 45°.

____

____

Name ______________________ Date __________ Class __________

LESSON 8-3

Practice C

Angle Relationships

Use the figure for Exercises 1–4.

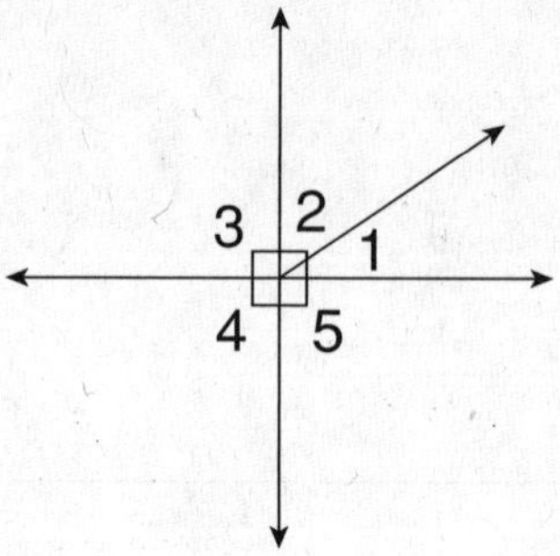

1. Name a pair of adjacent angles.

____

2. Name a pair of vertical angles.

____

3. Name a pair of complementary angles.

____

4. Name a pair of supplementary angles.

____

Find each unknown angle measure.

5. ∠1 and ∠2 are vertical angles. ∠1 measures 48°.

____

6. ∠3 and ∠4 are supplementary angles. ∠4 measures 117°.

____

7. ∠7 and ∠8 are complementary angles. ∠7 measures 23°.

____

8. ∠5 and ∠6 are congruent angles. ∠6 measures 12.5°.

____

9. If ∠1 and ∠2 are supplementary angles, and ∠3 and ∠2 are supplementary angles, what do you know about ∠1 and ∠3? Explain.

____

____

____

10. Two angles are complementary. One angle is 20° smaller than the other angle. What are the measures of both angles?

____

Name ______________________ Date ____________ Class ____________

LESSON 8-3
Reteach
Angle Relationships

When two lines intersect, two pairs of opposite angles, called vertical angles, are formed. Vertical angles are congruent.

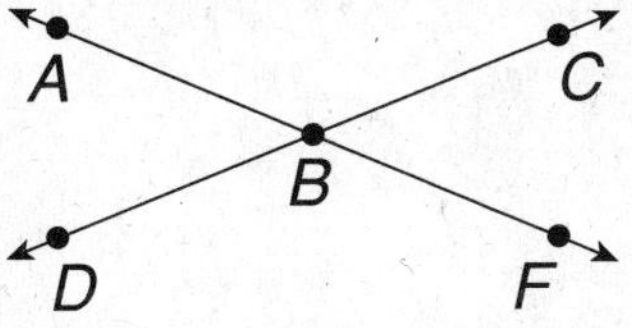

∠*ABC* and ∠*DBF* are vertical angles.
∠*ABD* and ∠*CBF* are vertical angles.

Adjacent angles have a common side and vertex.

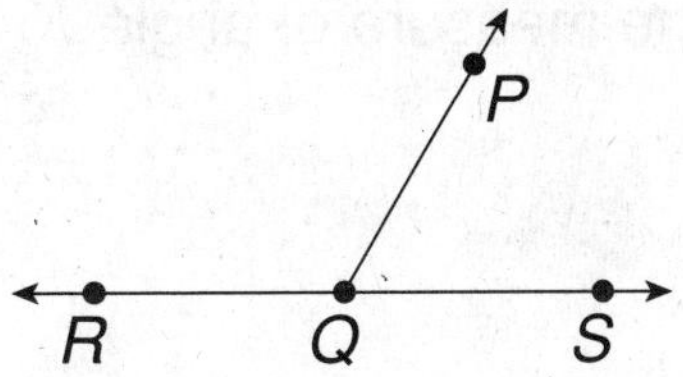

∠*PQR* and ∠*PQS* are adjacent angles.

Identify the type of each angle pair shown.

1.

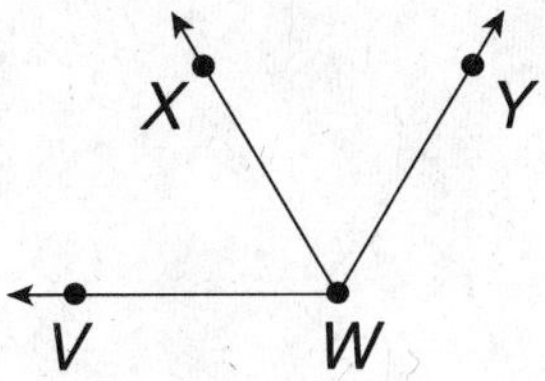

____

2.

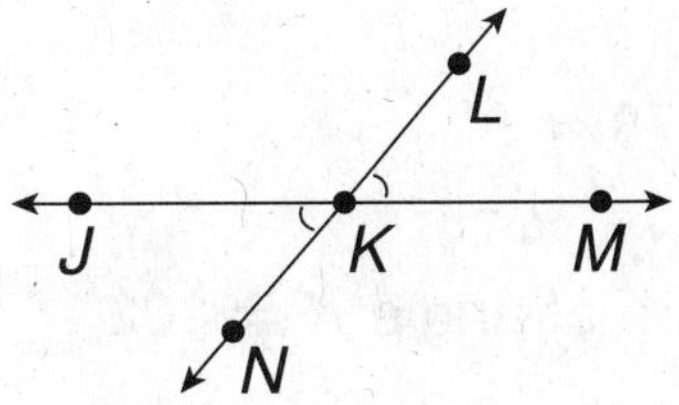

____

If the sum of the measures of two angles is 90°, then the angles are complementary angles.

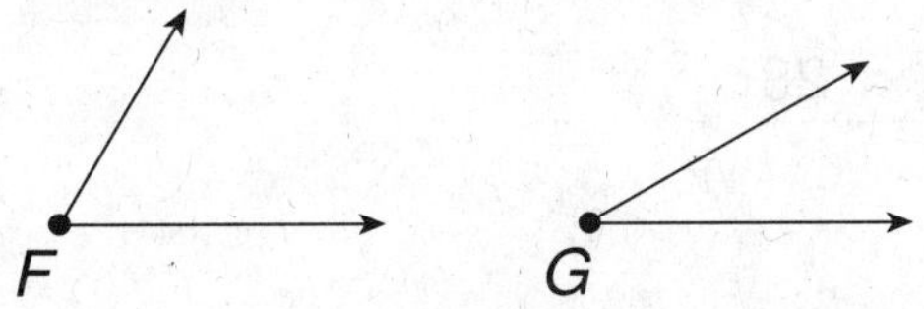

∠*F* and ∠*G* are complementary angles.

If the sum of the measures of two angles is 180°, then the angles are supplementary angles.

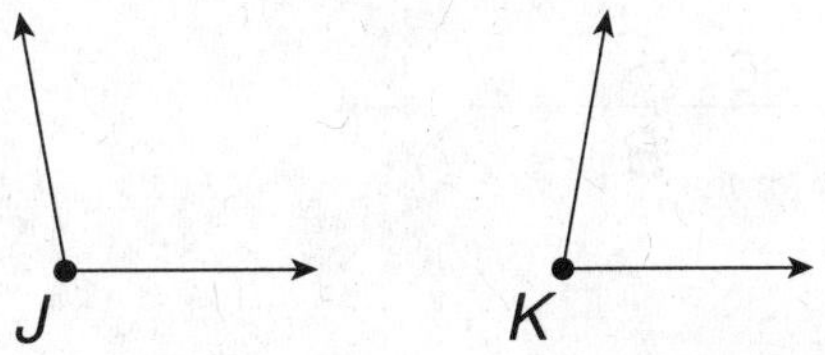

∠*J* and ∠*K* are supplementary angles.

Identify the type of each angle pair shown.

3.

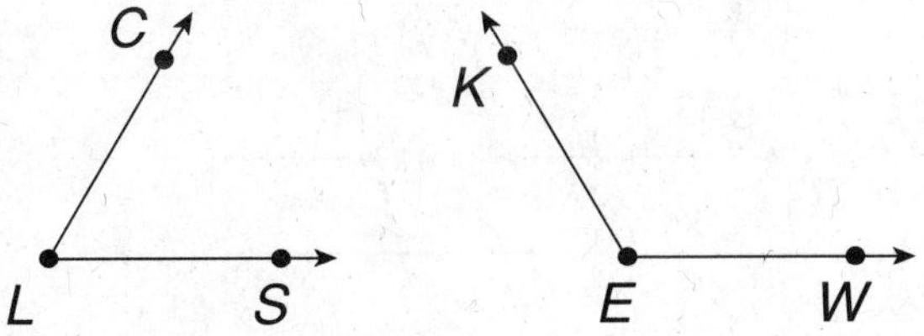

____

4.

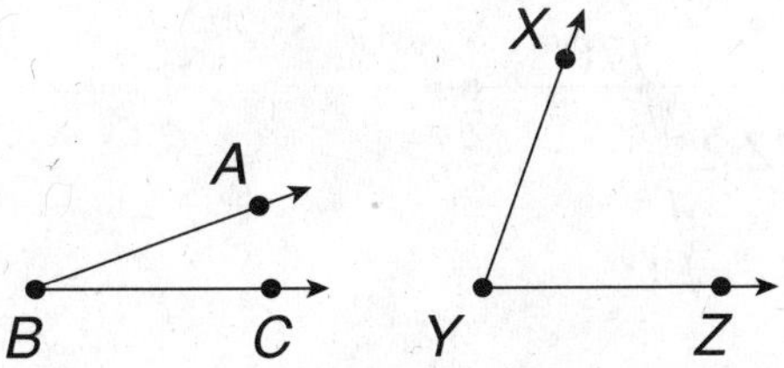

____

Name ______________________ Date ____________ Class ____________

LESSON 8-3

Reteach

Angle Relationships (continued)

You can use related facts to find an unknown angle measure.

If you know that angles *J* and *K* are supplementary and $m\angle J = 78°$, you can find the measure of angle *K*.

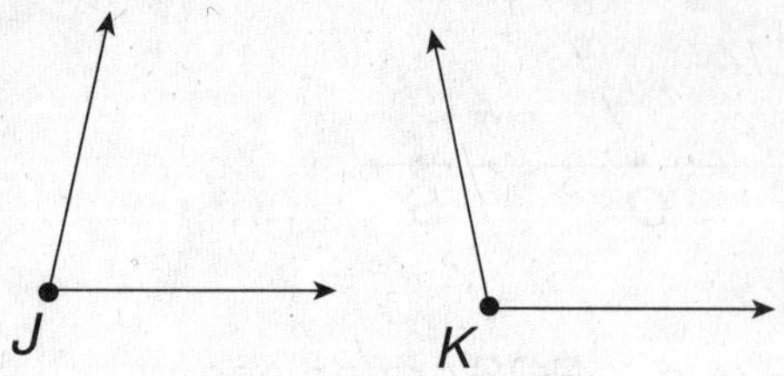

First write an equation.

$78 + K = 180$ Supplementary angles have a sum of 180°.

Then use related facts to solve the equation.

$78 + K = 180$
Think: $180 - 78 = K$
$K = 102$

So, the measure of angle $K = 102°$.

Find each unknown angle measure.

5. The angles are supplementary.

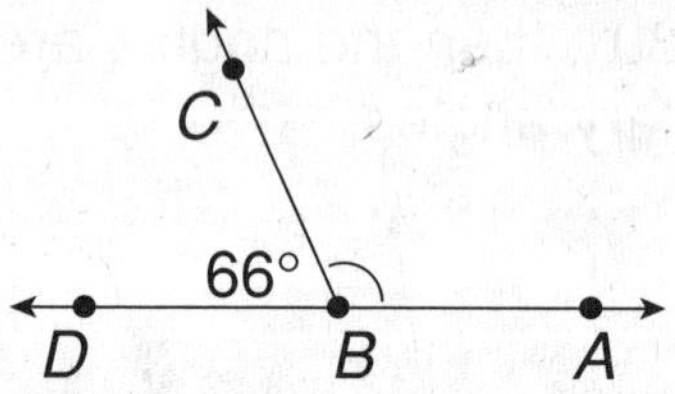

____

6. The angles are complementary.

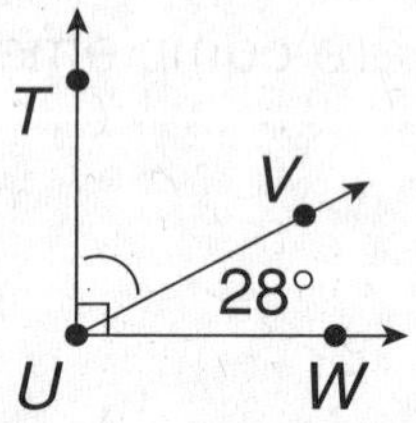

____

7. The angles are vertical angles.

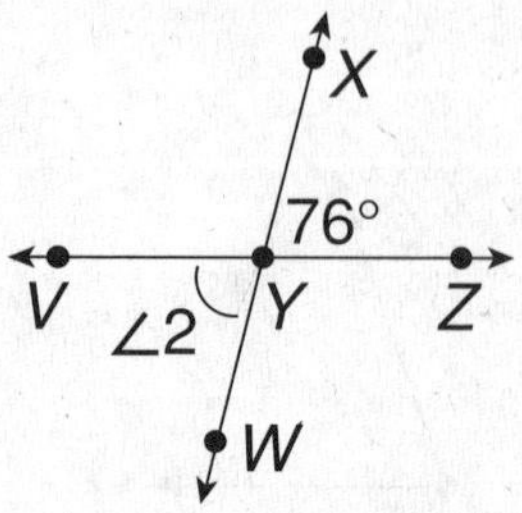

____

8. The angles are supplementary.

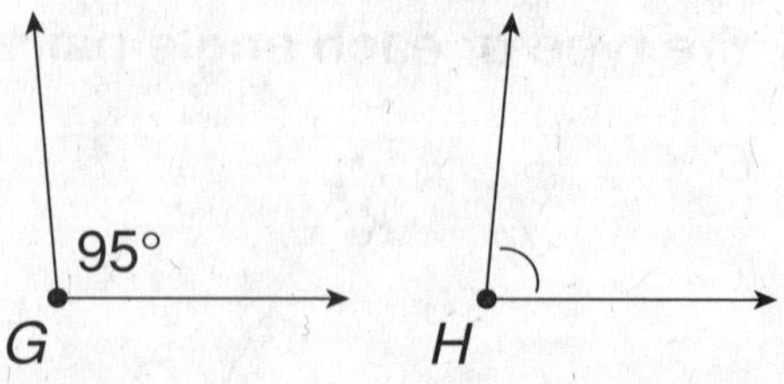

____

Name ______________________________ Date ____________ Class ____________

LESSON **8-3**

Challenge
Angle Partners

Measure all the angles in the chart below. Then draw lines to match each starting angle with its complementary and supplementary angles.

Complementary Angles	Starting Angles	Supplementary Angles
	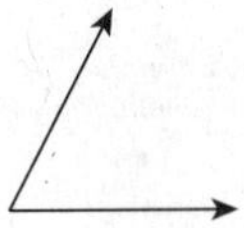	
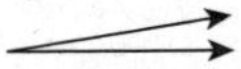	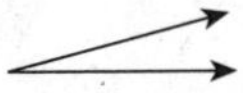	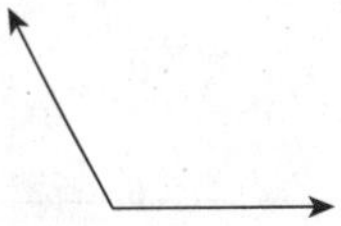
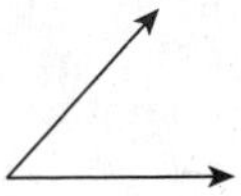	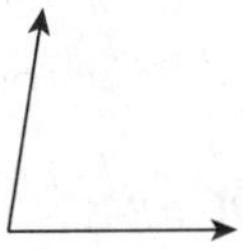	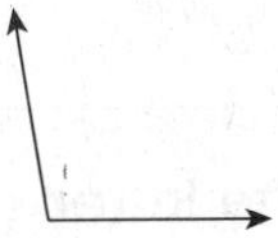
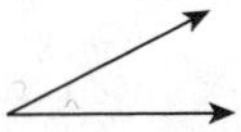	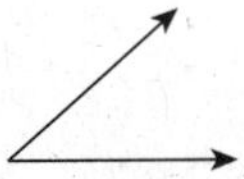	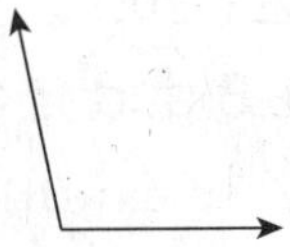
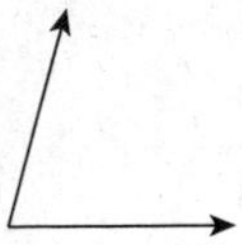	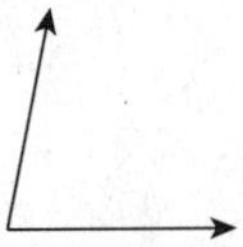	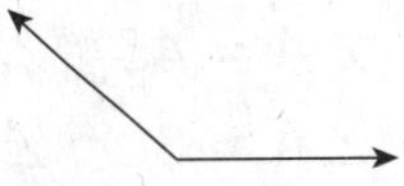

Name ______________________ Date ____________ Class ____________

LESSON 8-3

Problem Solving
Angle Relationships

Use the two compass roses to answer questions 1-6.

Cardinal Directions

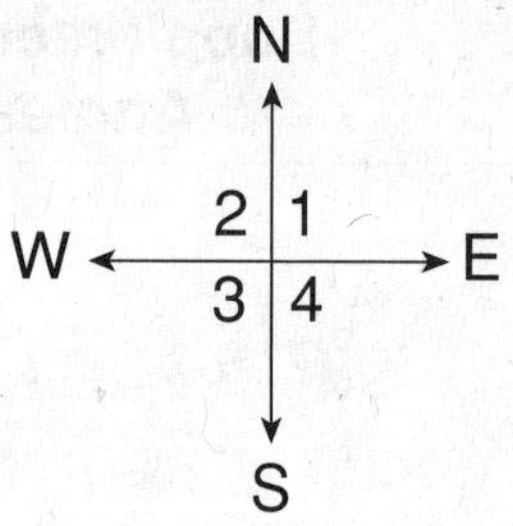

Intermediate Directions

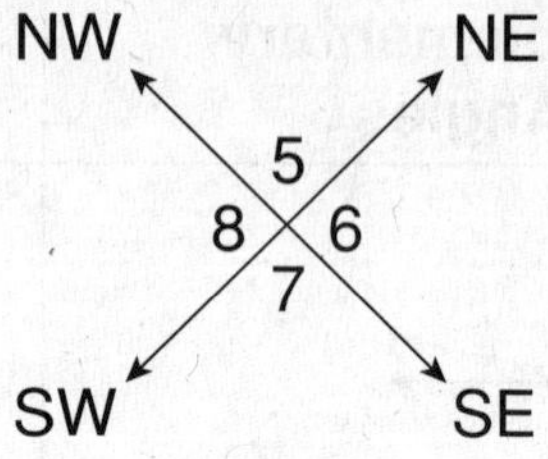

1. Which angles formed by the cardinal directions are vertical angles?

____

2. Which angles formed by the intermediate directions are vertical angles?

____

3. Draw the northwest directional ray on the cardinal compass rose. Describe the adjacent angles formed by the new ray.

____

4. North on a compass is 0°, and east is 90°. Use this information to label the degrees for each direction on the two compass roses above.

Circle the letter of the correct answer.

5. Which angles formed by the cardinal directions are supplementary to ∠2?

A ∠1

B ∠1 and ∠3

C ∠3 and ∠4

D ∠1, ∠3 and ∠4

6. Which angles formed by the intermediate directions are supplementary to ∠6?

F ∠5

G ∠5 and ∠7

H ∠7 and ∠8

J ∠5, ∠7 and ∠8

7. Angles *A* and *B* are complementary. ∠*B* is twice as large as ∠*A*. What are the measurements for each angle?

A ∠*A* = 45°; ∠*B* = 90°

B ∠*A* = 30°; ∠*B* = 60°

C ∠*A* = 60°; ∠*B* = 120°

D ∠*A* = 90°; ∠*B* = 180°

8. ∠1 and ∠2 are complementary. ∠2 and ∠3 are supplementary. The measure of ∠1 is 45°. What is the measure of ∠3?

F 45°

G 270°

H 90°

J 135°

Name ______________________ Date ____________ Class ____________

LESSON **8-3**

Reading Strategies

Draw Conclusions

Congruent means exactly the same size and shape. Angles that have the same measure are congruent.

Vertical angles are angles that are opposite each other. When two lines intersect, vertical angles are formed.

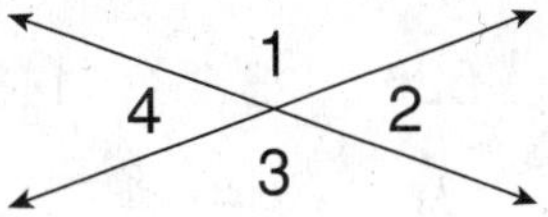

∠1 and ∠3 are vertical angles.

∠2 and ∠4 are vertical angles.

- Vertical angles have the same measure.
- Vertical angles are congruent.

Adjacent angles sit next to each other, or side-by-side. They share a common vertex and ray. Adjacent angles may or may not be congruent.

In the diagram above, ∠1 and ∠2 are adjacent angles. ∠4 and ∠3 are also adjacent angles.

Use this figure to answer the following questions. Use *congruent, vertical,* and *adjacent* to describe each angle. There may be more than one correct answer.

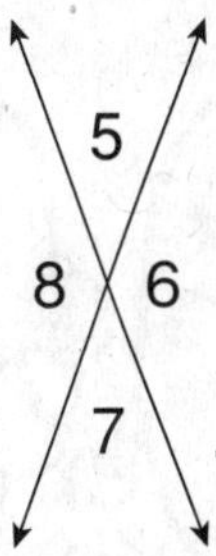

1. Study ∠5 and ∠6. What can you conclude about these two angles?

____

2. Study ∠5 and ∠7. What can you conclude about these two angles?

____

3. Study ∠6 and ∠8. What can you conclude about these two angles?

____

Name ______________________ Date __________ Class __________

LESSON **8-3**

Puzzles, Twisters & Teasers

Pick a Pair

Circle the pairs of angles that answer each question. Rearrange the letters from the unused angles to answer the riddle.

1. Which 2 pairs of angles are supplementary?

56°	92°	37°	118°	124°	62°	53°
K	**N**	**E**	**Y**	**G**	**Q**	**L**

2. Which 2 pairs of angles are vertical?

83°	42°	48°	126°	97°	126°	42°
J	**K**	**A**	**N**	**L**	**O**	**R**

3. Which 2 pairs of angles are complementary?

33°	109°	71°	76°	57°	104°	19°
W	**Y**	**P**	**E**	**V**	**B**	**H**

What kind of bean does not grow from seed? ______________

Name ______________________ Date ____________ Class ____________

LESSON 8-4

Practice A

Classifying Lines

Choose the correct answer.

1. Which relationship best describes the pair of lines shown below?

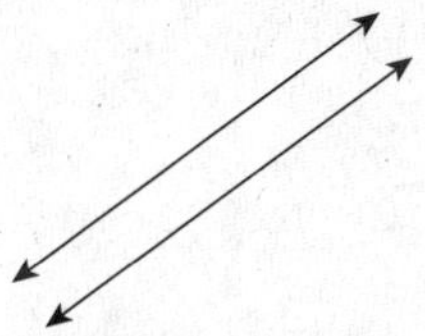

A They are parallel lines.
B They are intersecting lines.
C They are perpendicular lines.
D They are skew lines.

2. Which relationship best describes the pair of lines shown below?

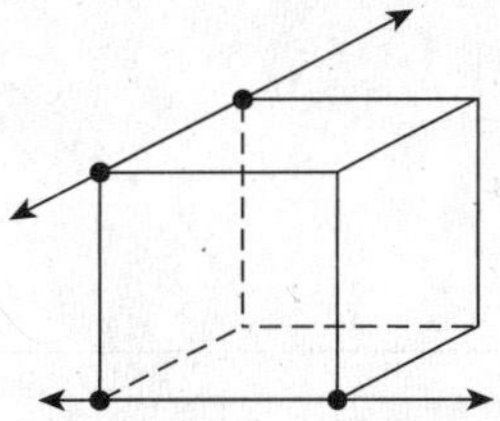

F They are parallel lines.
G They are intersecting lines.
H They are perpendicular lines.
J They are skew lines.

3. Which relationship best describes the pair of lines shown below?

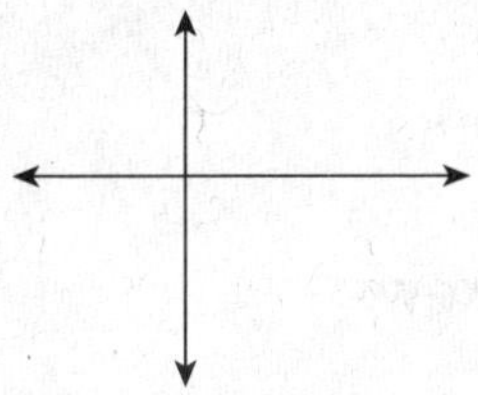

A They are parallel lines.
B They are intersecting lines.
C They are perpendicular lines.
D They are skew lines.

4. Which relationship best describes the pair of lines shown below?

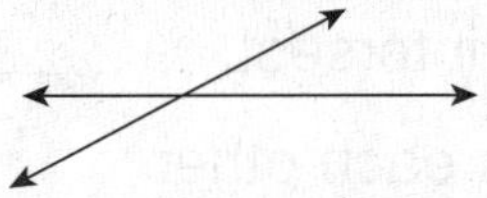

F They are parallel lines.
G They are intersecting lines.
H They are perpendicular lines.
J They are skew lines.

5. How is a pair of perpendicular lines different from a pair of intersecting lines?

____

____

6. What kinds of lines are similar to the rails of a railroad?

____

Name ______________________ Date ____________ Class ____________

LESSON 8-4

Practice B
Classifying Lines

Classify each pair of lines.

1.

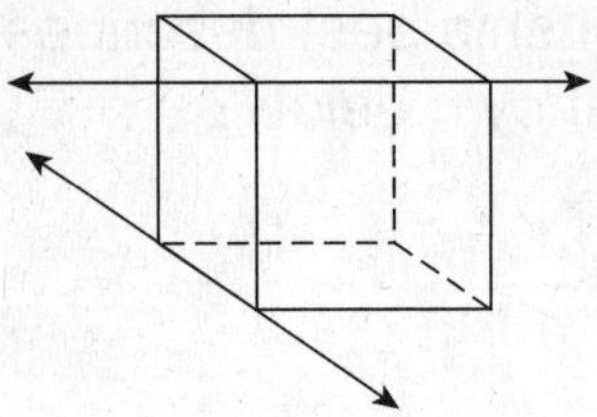

____

2.

____

3.

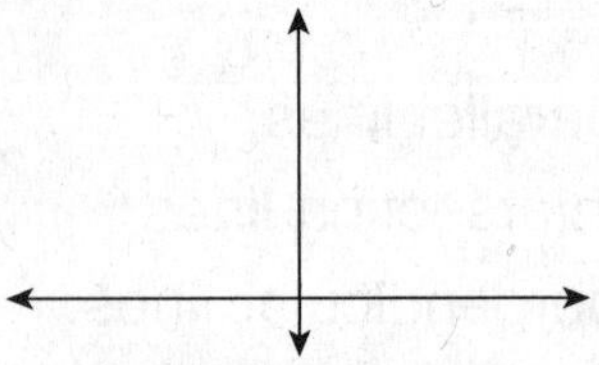

____

4.

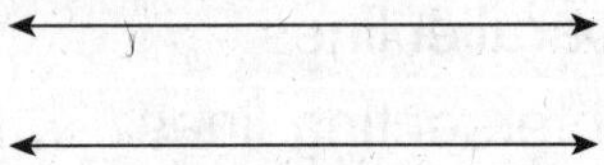

____

Match each description with its correct classification.

5. $\overleftrightarrow{AB}$ and $\overleftrightarrow{EF}$ lie on the same plane and never intersect. ______

6. $\overleftrightarrow{AB}$ and $\overleftrightarrow{EF}$ cross each other at one common point. ______

7. $\overleftrightarrow{AB}$ and $\overleftrightarrow{EF}$ lie on different planes and are neither parallel nor intersecting. ______

8. $\overleftrightarrow{AB}$ and $\overleftrightarrow{EF}$ intersect to form right angles. ______

A. $\overleftrightarrow{AB}$ intersects $\overleftrightarrow{EF}$.

B. $\overleftrightarrow{AB} \parallel \overleftrightarrow{EF}$

C. $\overleftrightarrow{AB}$ and $\overleftrightarrow{EF}$ are skew.

D. $\overleftrightarrow{AB} \perp \overleftrightarrow{EF}$

9. Oak Street runs parallel to Elm Street in a flat section of town. Tom tells you to meet him at the intersection of Oak and Elm. Explain why these instructions are impossible to follow.

____

____

10. Look around your classroom. Name a pair of parallel lines and a pair of perpendicular lines that you see.

____

____

Name ______________________ Date ____________ Class ____________

LESSON 8-4

Practice C
Classifying Lines

Use the diagram for Exercises 1–3.

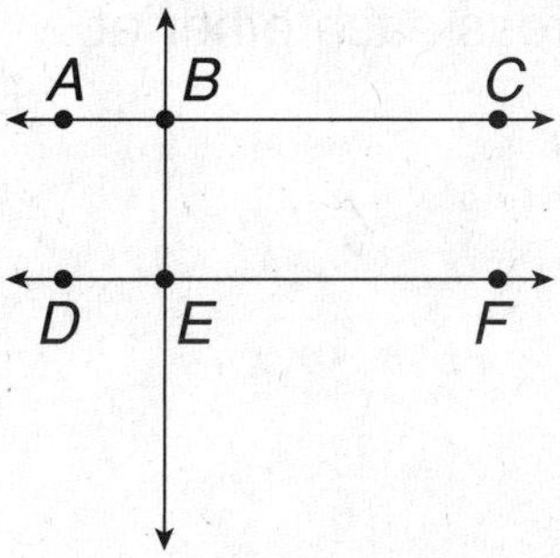

1. Name two pairs of perpendicular lines.

____

2. Name a pair of parallel lines.

____

3. Name two pairs of intersecting lines.

____

The lines in the figure intersect to form a rectangular box.

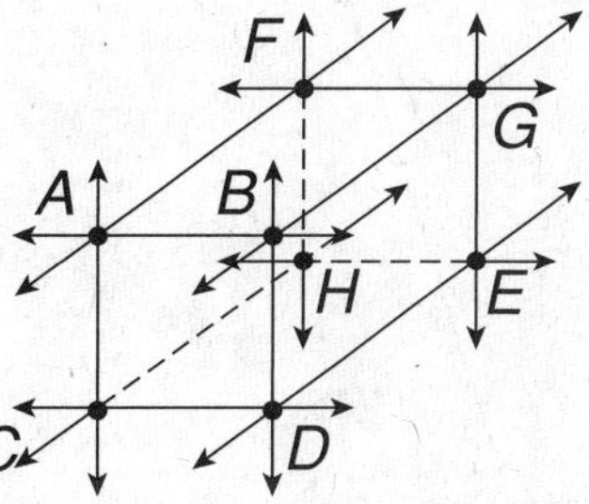

4. Name two pairs of lines that are parallel to $\overleftrightarrow{DE}$.

____

5. Name a pair of lines that are skew.

____

6. Name all the lines that are perpendicular to $\overleftrightarrow{AB}$.

____

7. Air traffic controllers must prevent two planes from sharing the same point in the sky at the same time or the planes will collide. What types of flight path lines will always prevent a collision of two planes?

____

8. Line *AD* is perpendicular to line *AB* at point *A*. Line *BC* is perpendicular to line *AB* at point *B*. All the lines lie in the same plane. What is the relationship between line *AD* and line *BC*?

____

Name ______________________ Date ____________ Class ____________

LESSON **8-4**

Reteach
Classifying Lines

Some lines have relationships.

Intersecting lines cross each other at one point.

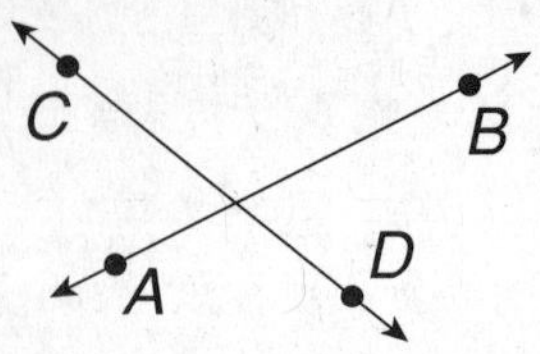

$\overleftrightarrow{AB}$ intersects $\overleftrightarrow{CD}$.

Perpendicular lines intersect to form right angles.

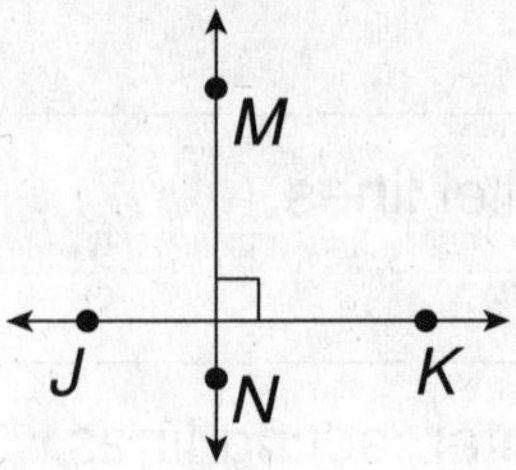

$\overleftrightarrow{JK} \perp \overleftrightarrow{MN}$

Parallel lines lie in the same plane but never intersect.

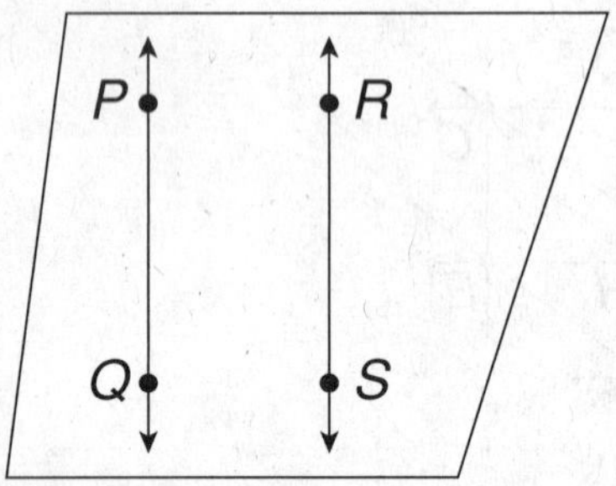

$\overleftrightarrow{PQ} \parallel \overleftrightarrow{RS}$

Skew lines lie in different planes and do not intersect.

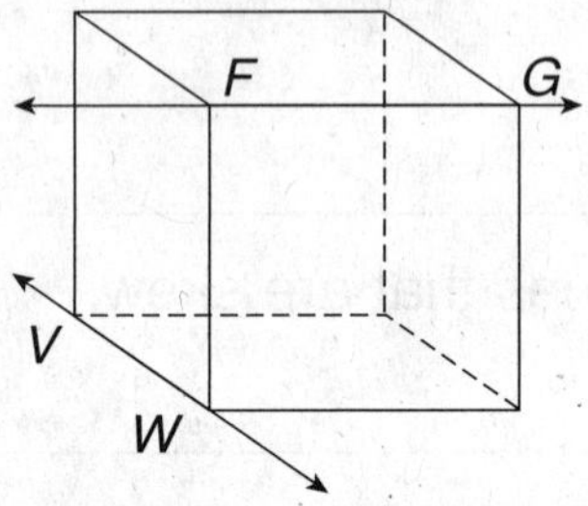

$\overleftrightarrow{FG}$ and $\overleftrightarrow{VW}$ are skew.

Classify each pair of lines.

1.

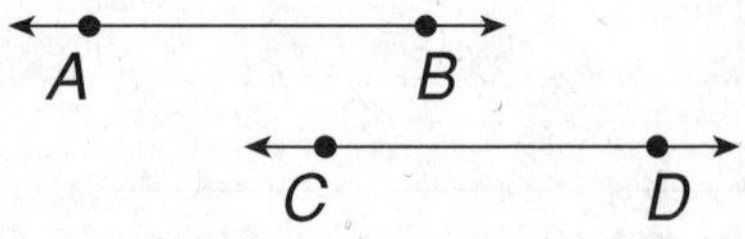

____

2.

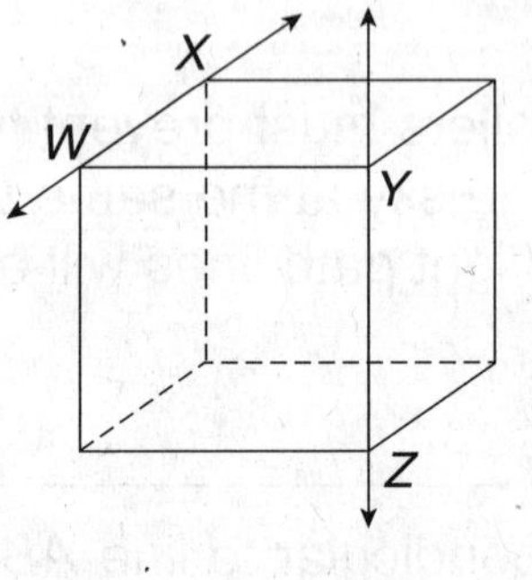

____

3.

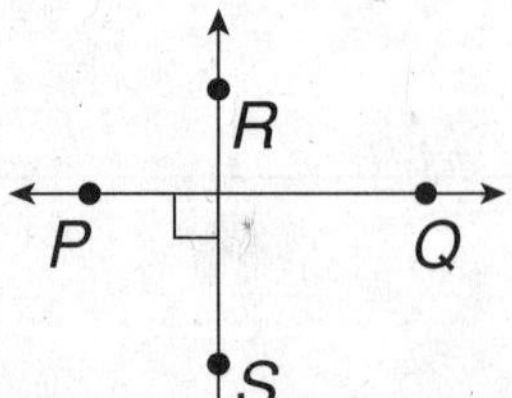

____

4.

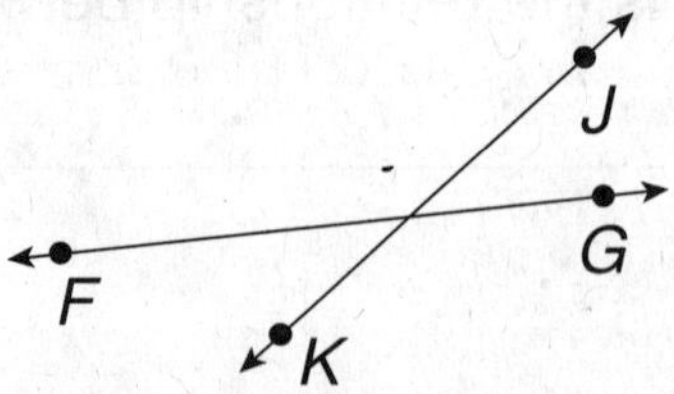

____

Name ______________________ Date __________ Class __________

LESSON 8-4

Challenge
Capital Streets

Washington, D.C., became the capital of the United States in 1800. The city's planners designed wide streets reaching out from two main centers—the White House and the Capitol Building. The President works and lives in the White House, and Congress works in the Capitol.

The map below shows the section of Washington, D.C, where the White House and Capitol are located. Use the following clues to label the streets on the map. Then name all the pairs of perpendicular streets on the map.

1. Madison Drive, Jefferson Drive, and Independence Avenue are parallel streets. Madison Drive is north of Jefferson Drive and Jefferson Drive is north of Independence Avenue.
2. 14th Street and 3rd Street are parallel streets. 14th Street is west of 3rd Street.
3. Pennsylvania Avenue intersects 14th Street near the White House.
4. The National Mall lies between Madison Drive to the north and Jefferson Drive to the south.

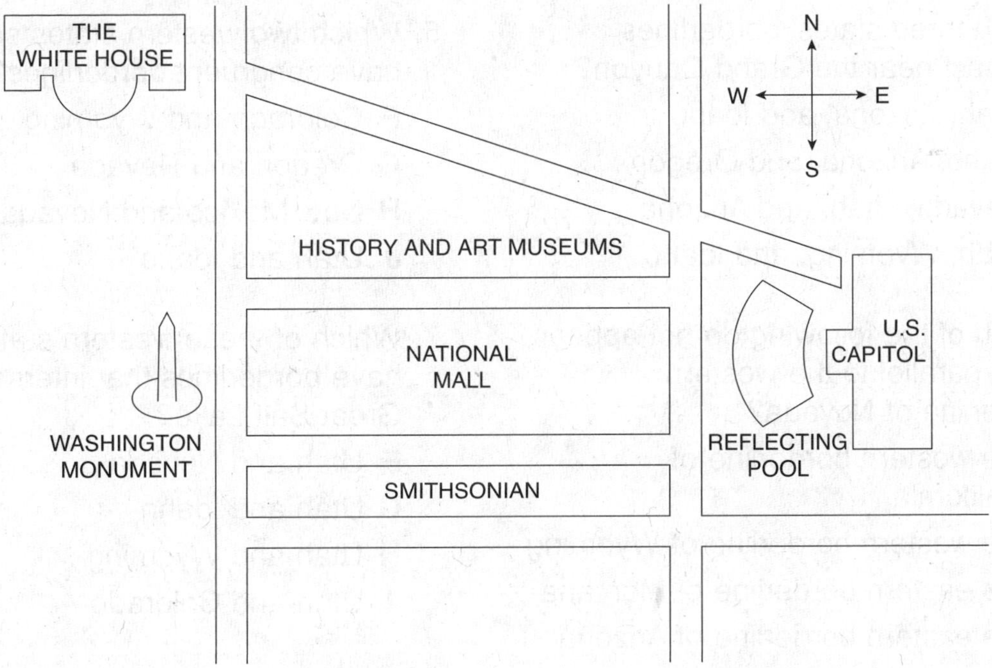

____

____

Name ______________________ Date ____________ Class ____________

LESSON 8-4

Problem Solving

Classifying Lines

Use the map to answer each question.

1. The area where the borders of Utah, Colorado, Arizona, and New Mexico meet is sometimes called the Four Corners. What kind of lines are formed where the borders meet?

____

2. Which borderlines on the map are skew lines?

____

3. What kinds of lines are suggested by the eastern and western borders of New Mexico?

____

Western U.S. States

Circle the letter of the correct answer.

4. Which three states' borderlines intersect near the Grand Canyon?

A Utah, Arizona, and Idaho
B Idaho, Arizona, and Oregon
C Nevada, Utah, and Arizona
D Utah, Wyoming, and Idaho

5. Which two western states seem to have congruent borderlines?

F Colorado and Wyoming
G Oregon and Nevada
H New Mexico and Nevada
J Utah and Idaho

6. Which of the following do not appear to be parallel to the western borderline of Nevada?

A the western borderline of California
B the western borderline of Wyoming
C the eastern borderline of Montana
D the eastern borderline of Arizona

7. Which of these western states do not have borderlines that intersect near Great Salt Lake?

F Utah and Nevada
G Utah and Idaho
H Utah and Wyoming
J Utah and Colorado

Name ______________________ Date __________ Class __________

LESSON **8-4**

Reading Strategies

Understanding Symbols

Parallel lines are the same distance apart and never meet.

The lines on a ruled piece of paper are similar to parallel lines.
The symbol || stands for parallel.

Segment *AB* is parallel to segment *CD*.

This can be written: *AB* || *CD*.

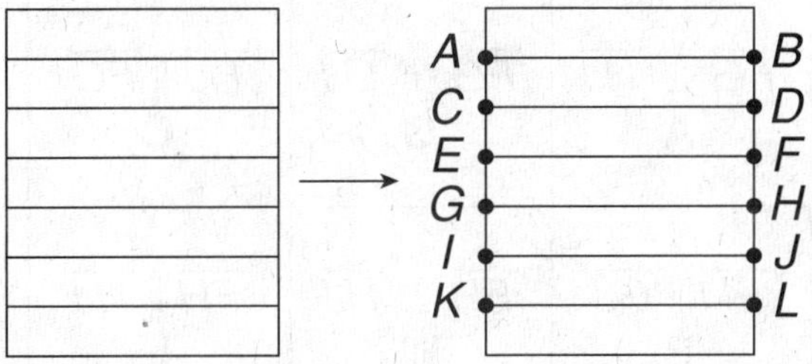

1. Find another pair of parallel line segments in the figure above.

____

2. Use the || symbol to write how the line segments are related.

____

Perpendicular lines meet to form square corners.

The symbol ⊥ stands for perpendicular. A square corner is used as a symbol to indicate that two lines are perpendicular to each other.

$\overline{TQ} \perp \overline{RS}$

Read: segment *TQ* perpendicular to segment *RS*.

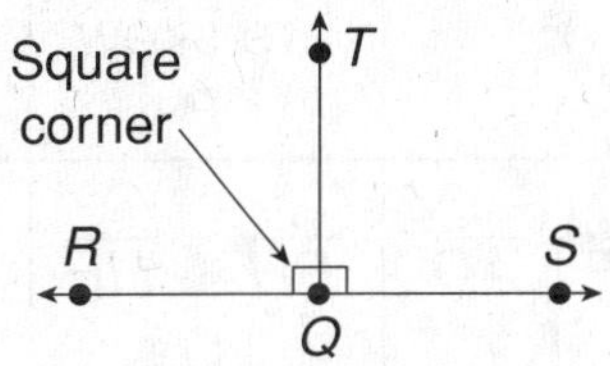

Use this figure to complete Exercises 3–5.

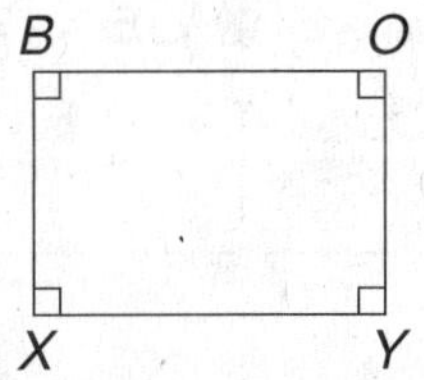

3. Segment *BX* is perpendicular to segment *BO*. Write this with symbols.

____

4. Find two other line segments that are perpendicular to each other.

____

5. What symbol in the figure tells you that line segments are perpendicular to each other?

____

Name ______________________ Date ____________ Class ____________

LESSON 8-4

Puzzles, Twisters & Teasers

What is my Line?

Use the figure on the right. For each box decide whether the statement is true or false. If it is false, black out the box.

$\overleftrightarrow{FG} \perp \overleftrightarrow{HJ}$ **A**	$\overleftrightarrow{FG} \perp \overleftrightarrow{GJ}$ **G**	$\overleftrightarrow{HJ} \parallel \overleftrightarrow{FG}$ **C**
$\overleftrightarrow{FH} \parallel \overleftrightarrow{GJ}$ **B**	$\overleftrightarrow{FH} \perp \overleftrightarrow{HJ}$ **W**	$\overleftrightarrow{FH} \parallel \overleftrightarrow{HJ}$ **L**
$\overleftrightarrow{FG} \parallel \overleftrightarrow{FH}$ **R**	$\overleftrightarrow{HJ} \perp \overleftrightarrow{GJ}$ **S**	$\overleftrightarrow{GJ} \parallel \overleftrightarrow{KL}$ **D**

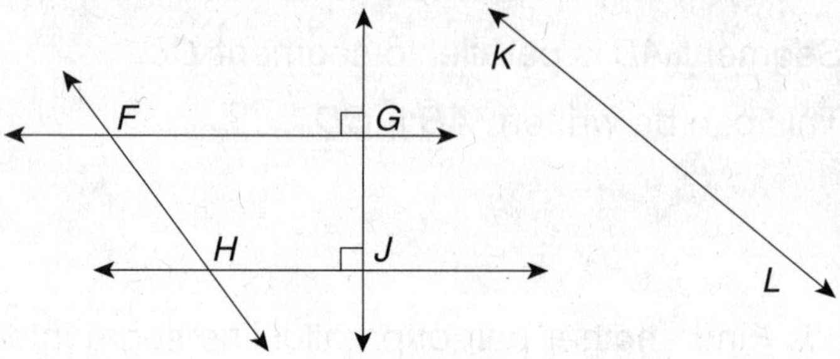

$\overleftrightarrow{AD} \perp \overleftrightarrow{AG}$ **E**	$\overleftrightarrow{BC}$ & $\overleftrightarrow{CD}$ are skew **P**	$\overleftrightarrow{DC} \parallel \overleftrightarrow{EF}$ **G**
$\overleftrightarrow{GF} \parallel \overleftrightarrow{AD}$ **I**	$\overleftrightarrow{BH} \perp \overleftrightarrow{HE}$ **O**	$\overleftrightarrow{BH} \parallel \overleftrightarrow{CE}$ **M**
$\overleftrightarrow{HG} \parallel \overleftrightarrow{AB}$ **O**	$\overleftrightarrow{HE} \perp \overleftrightarrow{CB}$ **F**	$\overleftrightarrow{HG}$ & $\overleftrightarrow{BC}$ are skew **N**

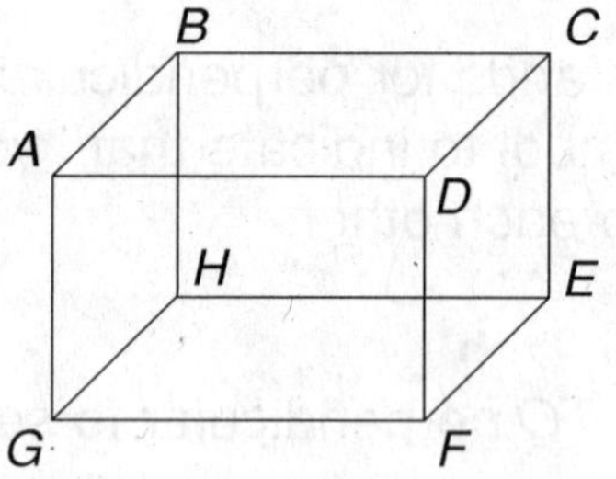

To solve the riddle, re-arrange the letters for the true boxes.

What do hens study in college? ______________________

Name ______________________ Date __________ Class __________

LESSON **8-5**

Practice A
Triangles

Classify each triangle as acute, obtuse, or right.

1.
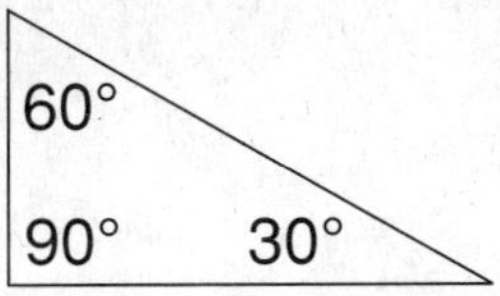

____

2.
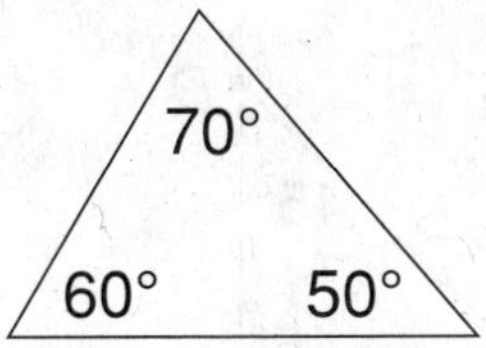

____

3.
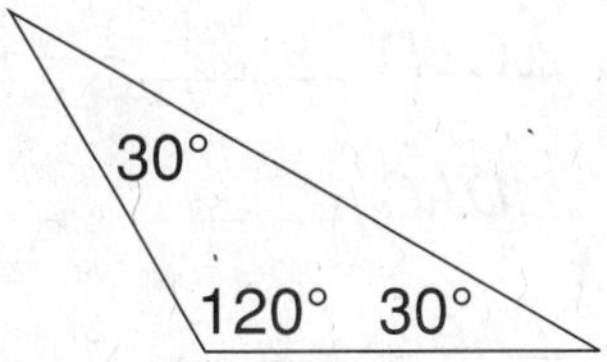

____

Classify each triangle as scalene, isosceles, or equilateral.

4.
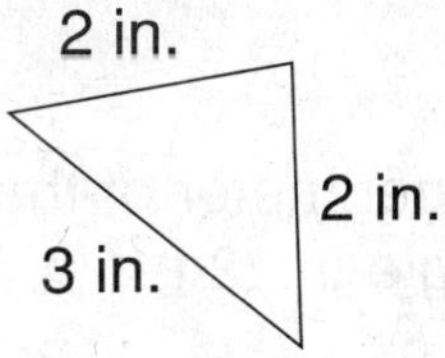

____

5.
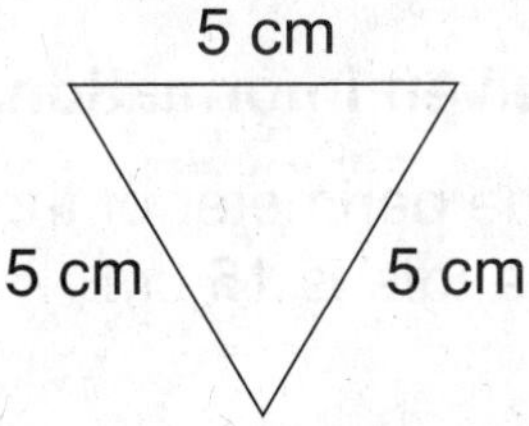

____

6.
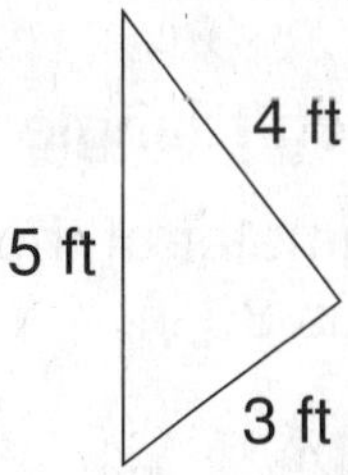

____

Find the measure of the unknown angle.

7.
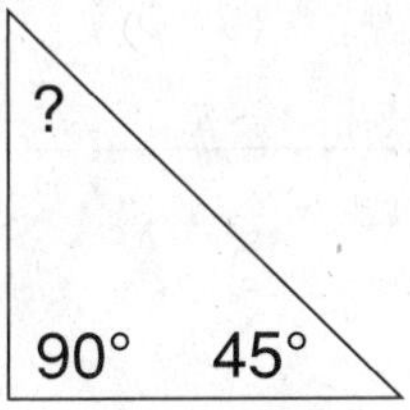

____

8.
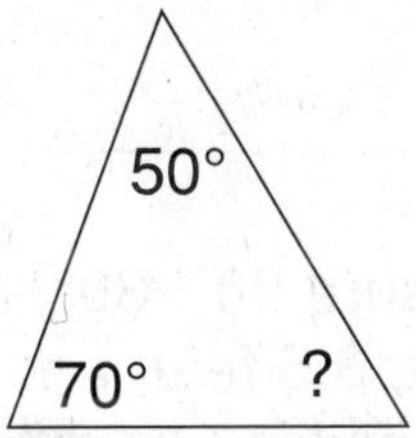

____

9.
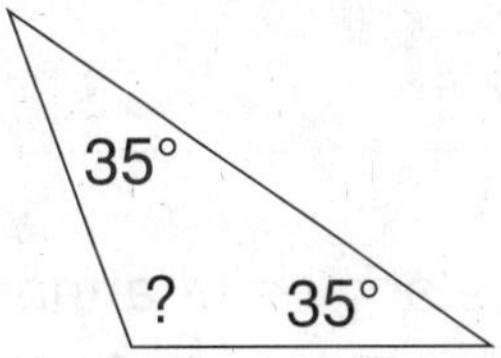

____

10. One side of an equilateral triangle measures 4 cm. What are the lengths of the other two sides of the triangle?

____

11. Karen says the angles of her triangle measure 90°, 50°, and 60°. Explain why this is impossible.

____

____

Name ______________________ Date ____________ Class ____________

LESSON 8-5

Practice B

Triangles

Use the diagram to find the measure of each indicated angle.

1. ∠*CBD* ______

2. ∠*DAC* ______

3. ∠*ABC* ______

4. ∠*EBA* ______

5. ∠*ACB* ______

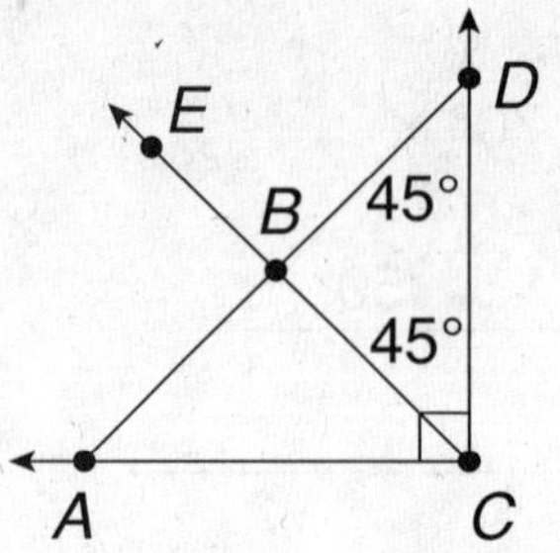

Classify each triangle using the given information.

6. The perimeter of the triangle is 30 in.

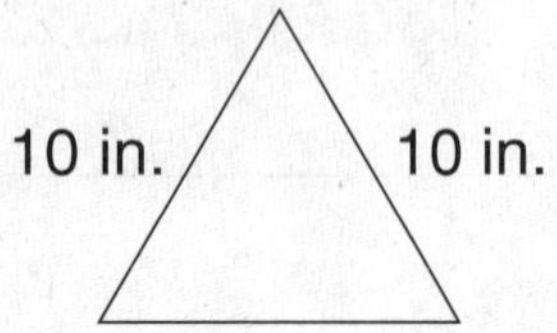

____

7. The perimeter of the triangle is 15 cm.

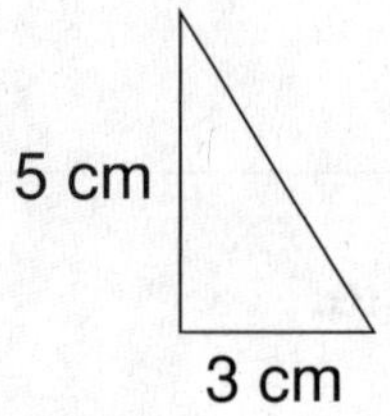

____

8. The perimeter of the triangle is 22 ft.

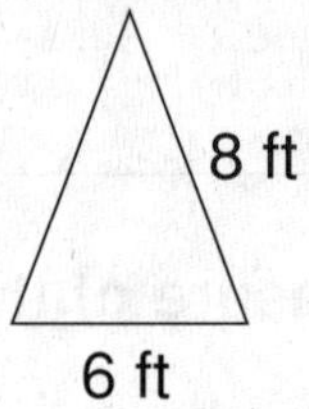

____

9. The angles of a triangular sail measure 90°, 30°, and 60°. Its sides measure approximately 2 feet, 3.5 feet, and 4 feet. Classify the triangular shape of the sail in two different ways.

____

10. Two angles in one triangle are congruent to two angles in another triangle. What can you conclude about the third angle in both triangles?

____

Name ______________________ Date ____________ Class ____________

LESSON 8-5

Practice C

Triangles

Use the diagram to find the measure of each indicated angle.

1. ∠*DEC* ______
2. ∠*ACB* ______
3. ∠*BCE* ______
4. ∠*EBC* ______

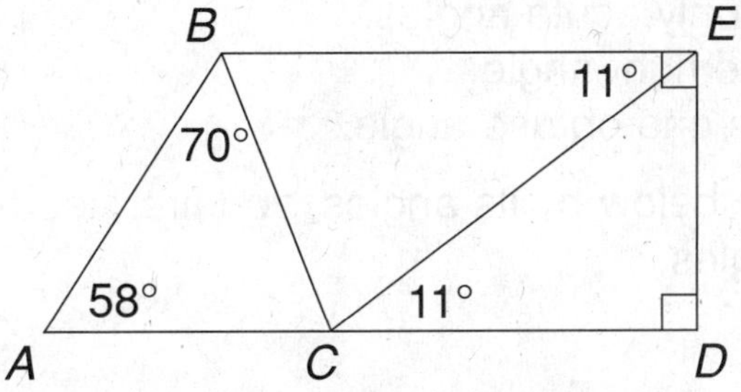

The lengths of sides $\overline{AB}$ and $\overline{BC}$ are given for △*ABC*. Use the perimeter to calculate the length of side $\overline{AC}$ and classify each triangle.

5. $AB = 1.5$; $BC = 4.8$;

   perimeter = 10

   ____

   ____

6. $AB = 2\frac{1}{4}$; $BC = 2\frac{1}{4}$;

   perimeter = $6\frac{3}{4}$

   ____

   ____

7. $AB = 638$; $BC = 181$;

   perimeter = 1,000

   ____

   ____

The measures of ∠*A* and ∠*B* are given for △*ABC*. Calculate the measure of ∠*C* and classify each triangle.

8. $\angle A = 22.17°$;

   $\angle B = 136.5°$

   ____

   ____

9. $\angle A = 60°$;

   $\angle B = \frac{1}{2} \cdot \angle A$

   ____

   ____

10. $\angle A = 26.5°$;

    $\angle B = 3 \cdot \angle A$

    ____

    ____

11. Starting from the airport, a pilot flew directly north for 112 miles. Then she made a 90° turn and flew east for 84 miles to reach Bear Lake. If she flew in a straight line, in which direction did the pilot fly to get back to the airport? If she flew a total of 336 miles, how far is the airport from Bear Lake?

    ____

12. ∠*A* of △*ABC* is congruent to ∠*G* of △*FGH*. ∠*B* of △*ABC* and ∠*F* of △*FGH* are supplementary. ∠*C* measures 29°, and ∠*A* measures 37°. What is the measure of each angle in △*FGH*?

    ____

Name ______________________ Date __________ Class __________

LESSON 8-5

Reteach
Triangles

You can classify triangles by the measures of their angles.

An acute triangle has only acute angles.
A right triangle has one right angle.
An obtuse triangle has one obtuse angle.

To classify the triangle below by its angles, you first need to find the measure of all the angles.

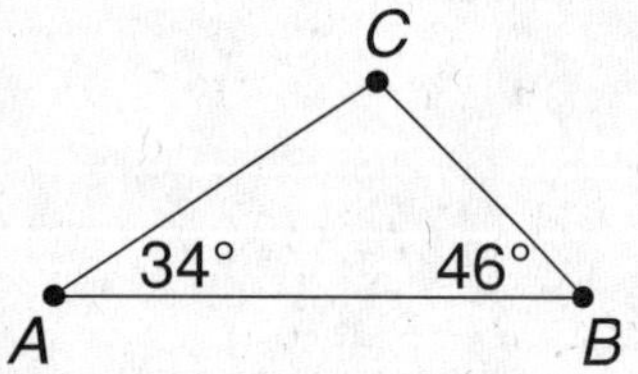

The sum of the angles of a triangle is 180°.

To find the unknown angle, first find the sum of the two known angles.

34 + 46 = 80

Then subtract the sum from 180.

180 − 80 = 100

The difference is the measure of the third angle.

The measures of the angles are 34°, 46°, and 100°. So the triangle is obtuse.

Find each unknown angle. Then classify each triangle.

1.

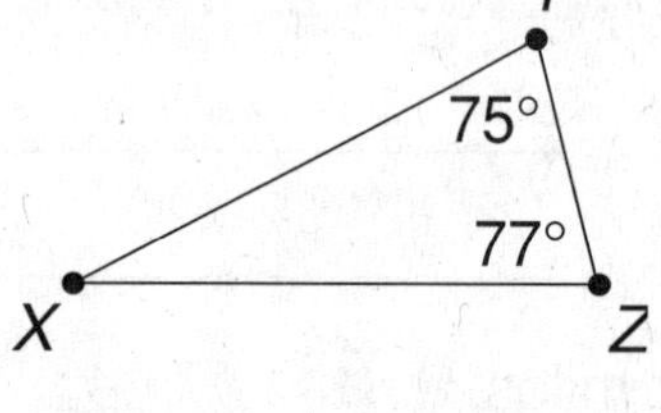

____

2.

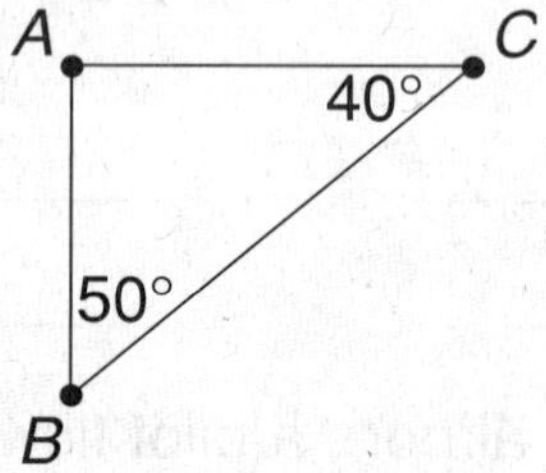

____

3.

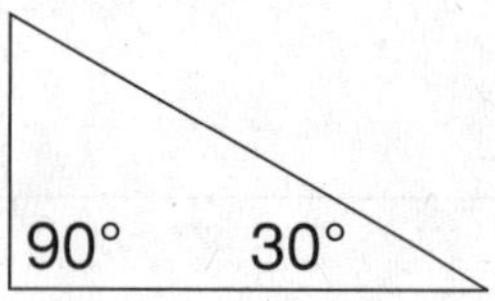

____

4.

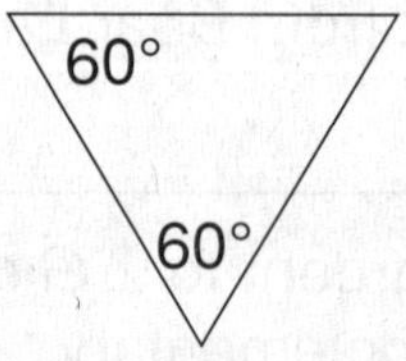

____

Name ______________________ Date __________ Class __________

LESSON 8-5

Reteach

Triangles (continued)

You can also classify triangles by the lengths of their sides.

A scalene triangle has no congruent sides.
An isosceles triangle has at least 2 congruent sides.
An equilateral triangle has 3 congruent sides.

The perimeter of the triangle below is 18.6 cm. To classify the triangle below by its sides, you first need to find the length of the unknown side.

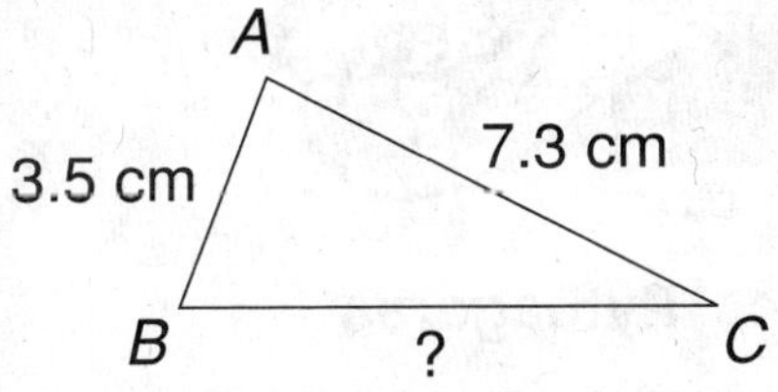

To find the unknown length, first find the sum of the two known lengths.

$7.3 + 3.5 = 10.8$

Then subtract this sum from the perimeter.

$18.6 - 10.8 = 7.8$

The length of the third side is 7.8 cm.

The lengths of the sides of the triangle are 3.5 cm, 7.3 cm, and 7.8 cm, so the triangle is scalene.

Find each missing length. Then classify each triangle.

3.

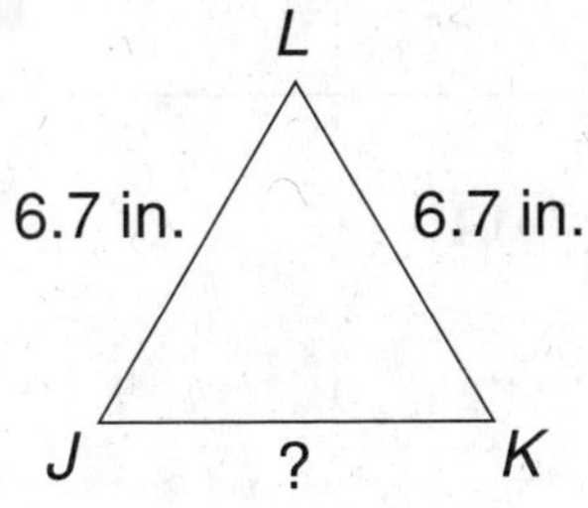

The perimeter of the triangle is 20.1 inches.

____

4.

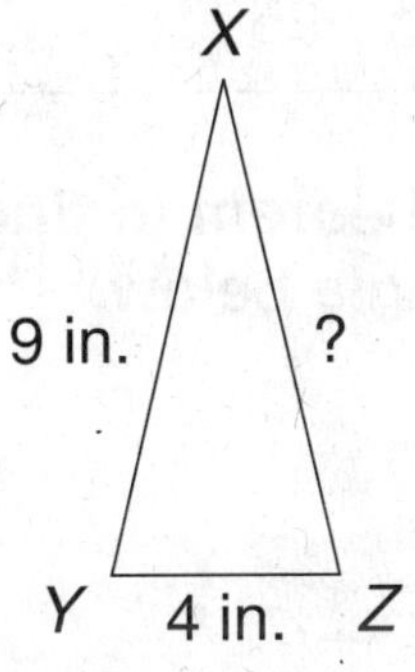

The perimeter of the triangle is 22 inches.

____

Name ______________________ Date __________ Class __________

LESSON 8-5

Challenge

Square Legs

About 2,400 years ago a Greek philosopher and mathematician named Pythagoras proved a very important rule for triangles. Today people all over the world study and use this rule named in his honor—the Pythagorean Theorem. It states the relationship between the side lengths of a right triangle.

Pythagoras

PYTHAGOREAN THEOREM

In a right triangle, the square of the length of the hypotenuse is equal to the sum of the squares of the lengths of the legs.

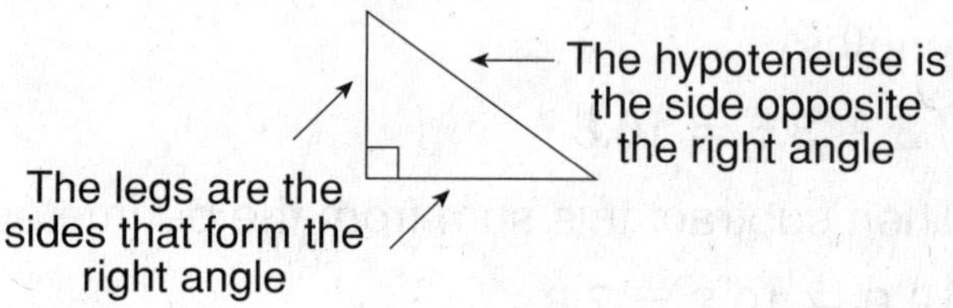

$$(\text{leg})^2 + (\text{leg})^2 = (\text{hypotenuse})^2$$

$$a^2 + b^2 = c^2$$

$$3^2 + 4^2 = 5^2$$

$$9 + 16 = 25$$

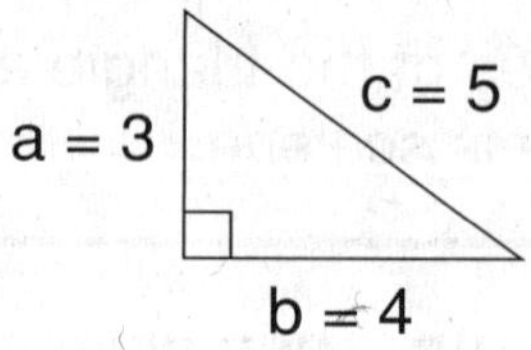

Use the Pythagorean Theorem to find the length of the missing side for each right triangle below.

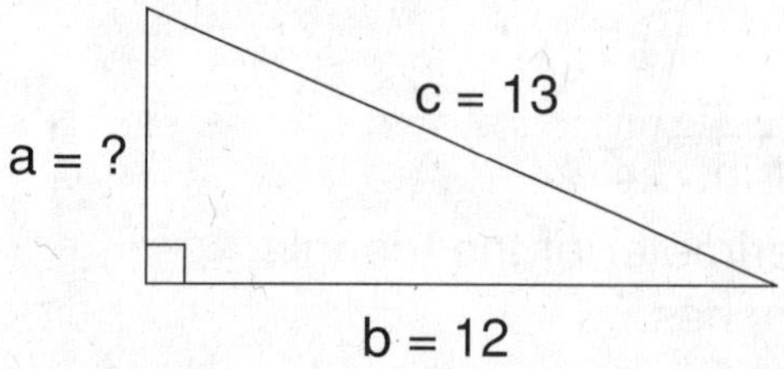

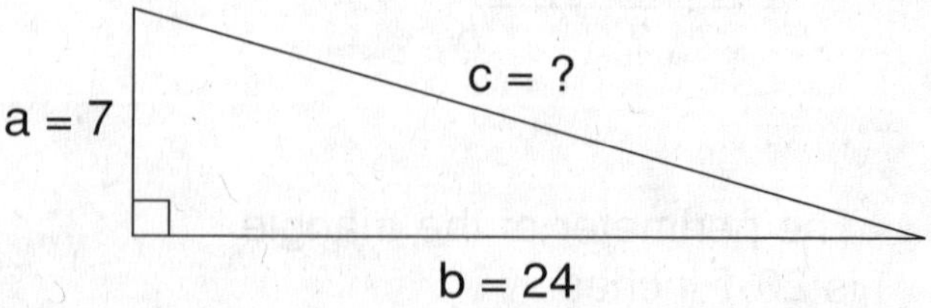

__________________ __________________

Name ______________________ Date ____________ Class ____________

LESSON **8-5**

Problem Solving

Triangles

Use the triangle diagram to answer each question.

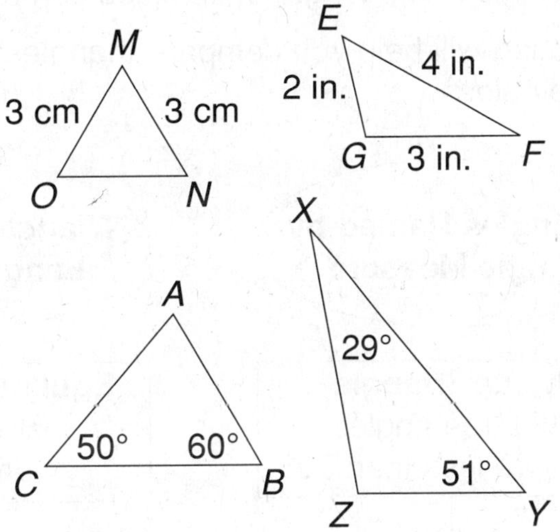

1. Classify triangle *ABC*. What is the measure of the missing angle?

____

2. Classify triangle *XYZ*. What is the measure of the missing angle?

____

3. If triangle *MNO* is an equilateral triangle, what is the measure of the missing side?

____

Circle the letter of the correct answer.

4. What is the complement of ∠*XYZ*?

A 39°

B 51°

C 129°

D 309°

5. Classify triangle *EFG*.

F scalene triangle

G isosceles triangle

H equilateral triangle

J right triangle

6. Which of the following statements is always true?

A A right triangle is a scalene triangle.

B An equilateral triangle is an isosceles triangle.

C An isosceles triangle is an obtuse triangle.

D A right triangle is an acute triangle.

7. Which of the following is not true of all right triangles?

F The sum of the measures of the angles is 180°.

G Two of its angles are supplementary angles.

H At least two of its angles are acute.

J The side with the greatest length is opposite the right angle.

Name ______________________ Date ____________ Class ____________

LESSON **8-5**

Reading Strategies

Graphic Organizer

A triangle can be classified by the measurement of one or more of its angles. The length of its sides can also classify a triangle.

This chart will help you compare triangles by their angles and by their sides.

Triangles Named by Angle Measure

Acute Triangle
All three angles measure less than 90°.

↓

Right Triangle
One angle measures 90°.

↓

Obtuse Triangle
One angle measures more than 90°.

Triangles Named by Length of Sides

Equilateral Triangle
All the sides are equal.

↓

Isosceles Triangle
Two sides are equal.

↓

Scalene Triangle
None of the sides are equal.

Write two of the following terms to describe each triangle:
acute, right, obtuse, equilateral, isosceles, scalene.

1.

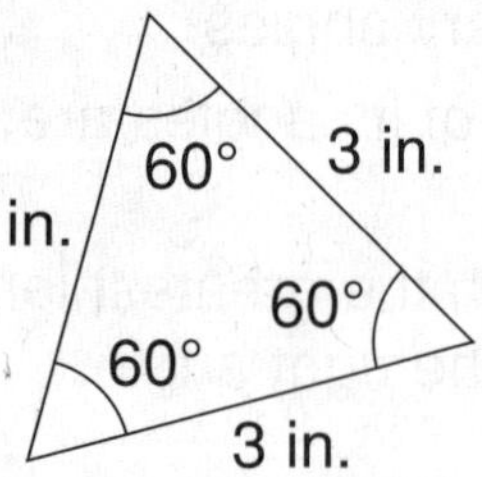

____

2.

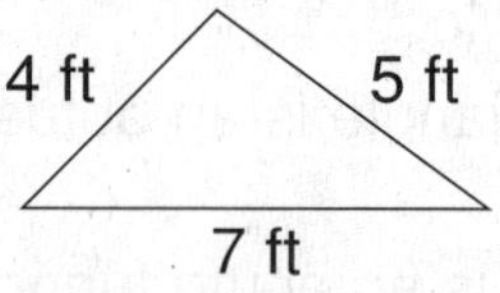

____

Name ______________________ Date ____________ Class ____________

LESSON 8-5

Puzzles, Twisters & Teasers

Find the Measure

Find the missing angle measure in each triangle. When you have completed the problem, look at the blank lines at the bottom of the page. Notice if any of the angle degrees of the letters match the degrees under the blank lines. If they do, write the letter of the angle on the line. The resulting message will be the answer to the riddle.

1.

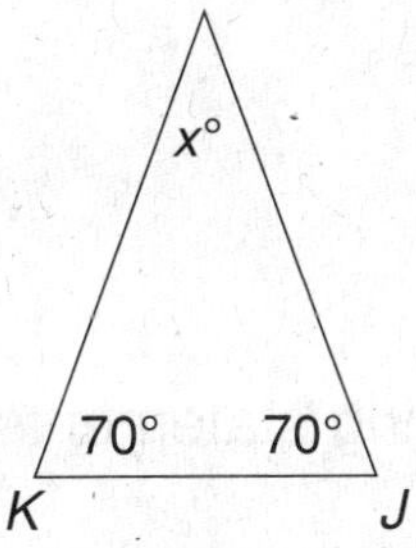

2.

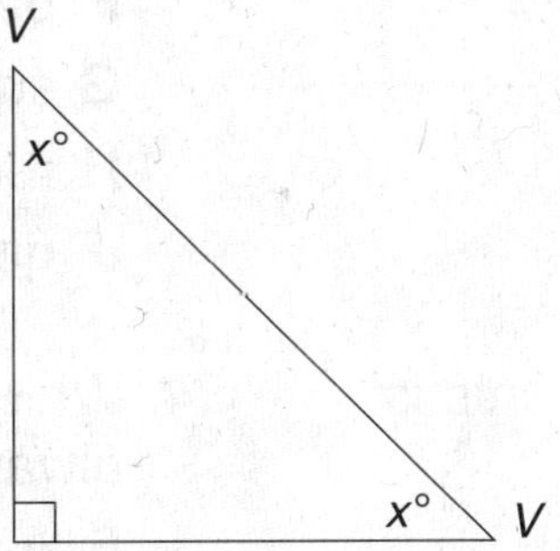

3.

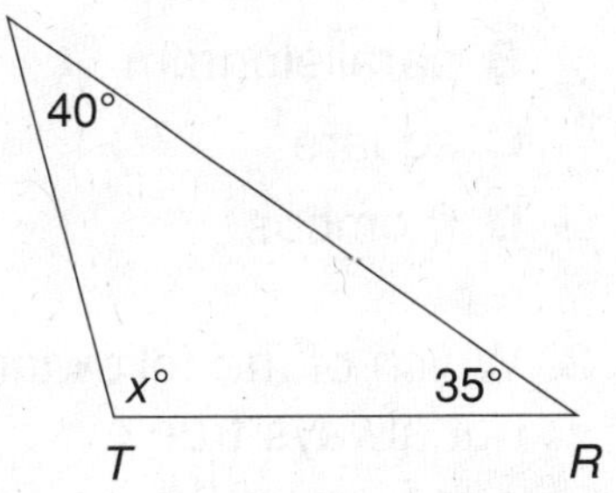

4.

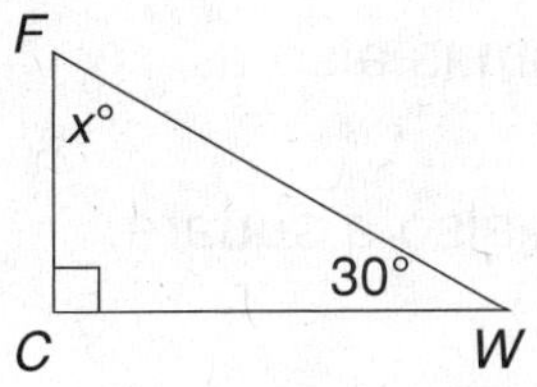

5.

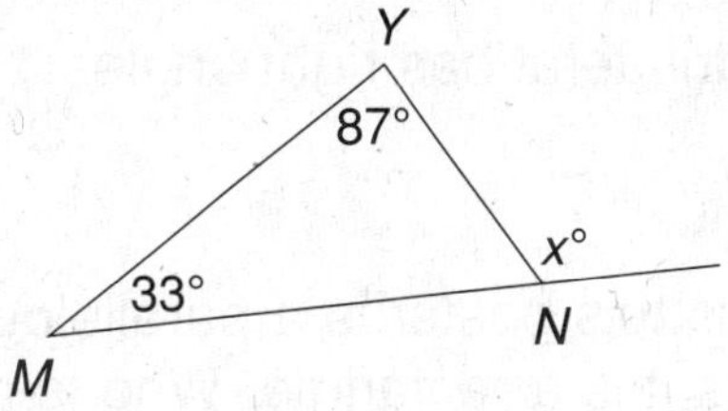

6.

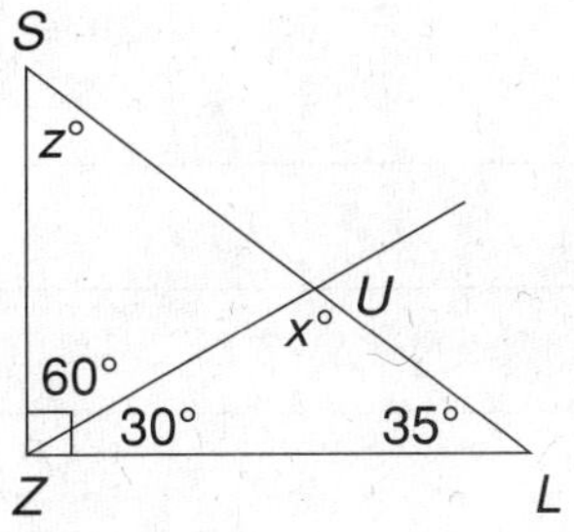

7.

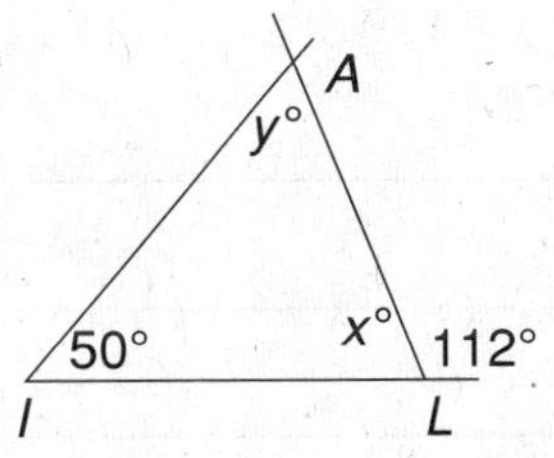

Why do you like studying about volcanoes?

___	___ ,	___	___	___	___	___	___	___	___	___
50°	105°	55°	62°	68°	62°	45°	62°	60°	115°	120°

Name ______________________ Date __________ Class __________

LESSON 8-6

Practice A

Quadrilaterals

Choose the correct answer.

1. Which name is most descriptive of the figure shown below?

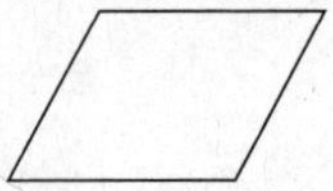

A quadrilateral
B parallelogram
C square
D rhombus

2. Which name is most descriptive of the figure shown below?

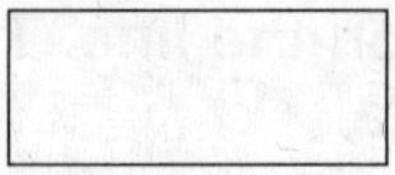

F parallelogram
G square
H rectangle
J quadrilateral

3. Which of the following statements is not always true?

A A quadrilateral has 4 sides.
B A quadrilateral has 4 angles.
C A quadrilateral has straight sides.
D A quadrilateral has right angles.

4. Which of the following statements is always true?

F A trapezoid is also a parallelogram.
G A square is also a rhombus.
H A parallelogram is also a rectangle.
J A rectangle is also a square.

5. Carol says this poster is a parallelogram. Anita says it is a rectangle. Who is right? Explain.

____

____

____

6. Joel hung two square mirrored tiles side-by-side in his bathroom. What is the best description for the geometric figure the two mirrors formed together?

____

Name ______________________ Date ____________ Class ____________

LESSON 8-6

Practice B
Quadrilaterals

Give the most descriptive name for each figure.

1. 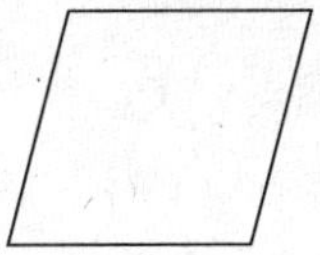

____

2. 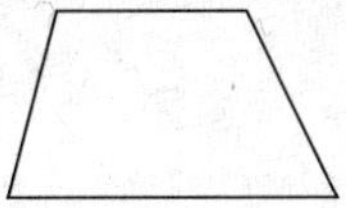

____

3. 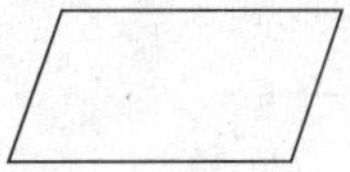

____

4. 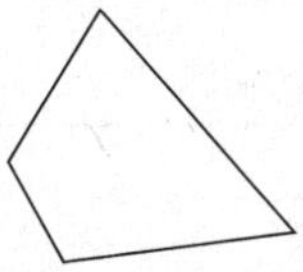

____

5. 

____

6. 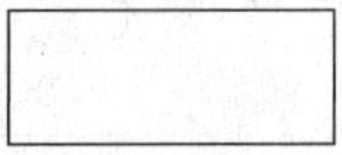

____

Complete each statement.

7. All rectangles are also ______________.

8. A rhombus is sometimes a ______________.

9. All trapezoids are also ______________.

10. A ______________ is any plane figure with four straight sides and four angles.

11. A quadrilateral with two sets of parallel lines, but does not have 90° angles is called a ______________.

12. Devon made a table top in the shape of a quadrilateral. All of its angles measure 90°. What could the shape of Devon's table top be?

____

13. The perimeter of a rhombus is 64 inches. What is the length of each side of the rhombus? Explain.

____

14. Explain why a trapezoid is a quadrilateral, but a quadrilateral is not always a trapezoid.

____

____

____

Name ______________________ Date __________ Class __________

LESSON **8-6**

Practice C

Quadrilaterals

Give all the possible names for each figure. Circle the most exact name.

1. 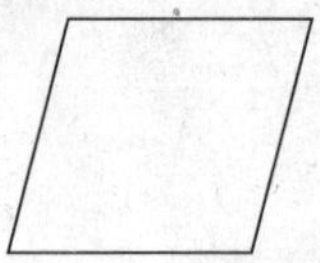

____

____

2. 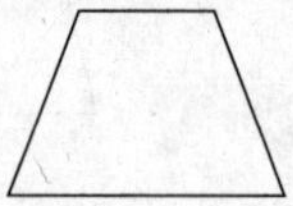

____

____

3. 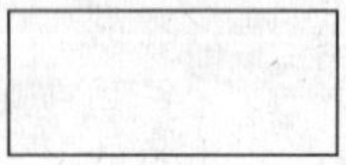

____

____

4. 

____

____

Draw each quadrilateral as it is described. If it is not possible, explain.

5. a parallelogram that is also a trapezoid

____

____

6. a rhombus that is not a parallelogram

____

____

7. Draw an example of a quadrilateral that is not a parallelogram or a trapezoid.

8. All the sides of a quadrilateral measure 25 cm. Explain why this is not enough information to classify what kind of quadrilateral it is.

____

____

Name ______________________ Date __________ Class __________

LESSON 8-6

Reteach

Quadrilaterals

A quadrilateral is a plane figure with four sides and four angles.

There are special types of quadrilaterals.

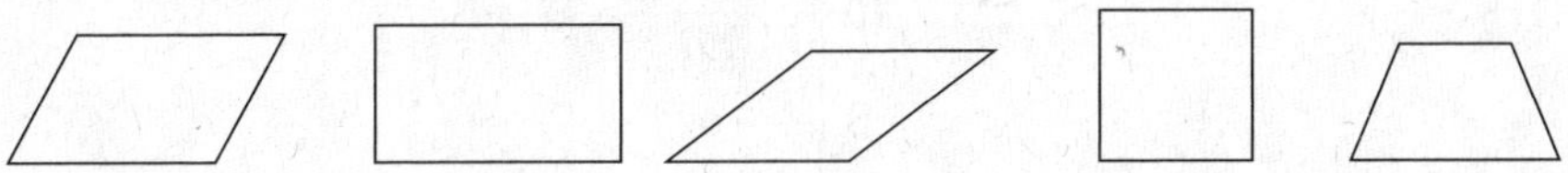

parallelogram rectangle rhombus square trapezoid

A parallelogram has opposite sides that are parallel and congruent. Opposite angles are also congruent.

A rectangle is a parallelogram that has four right angles.

A rhombus is a parallelogram with four congruent sides.

A square is a rectangle with four congruent sides.

A trapezoid is a quadrilateral with exactly two parallel sides.

The best description for the figure below is rectangle because it is a parallelogram with four right angles.

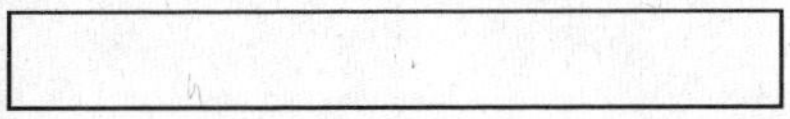

Give the best name for each figure.

1. 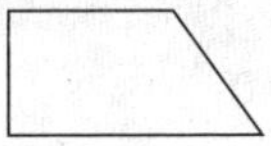

____

2. 

____

3. 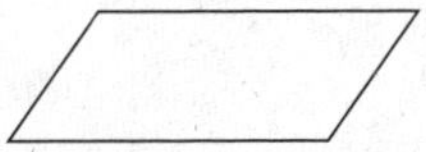

____

4. 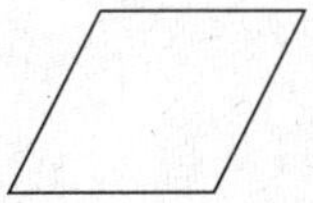

____

5. 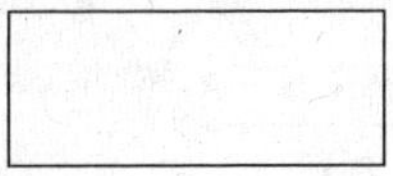

____

6. 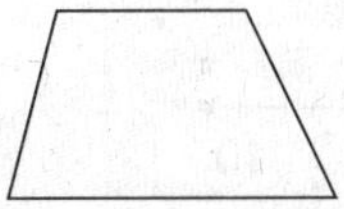

____

Name ______________________ Date __________ Class __________

LESSON 8-6

Challenge

What's the Point?

For each grid below, name the ordered pair of the missing fourth point that could be used to complete the given quadrilateral.

1.

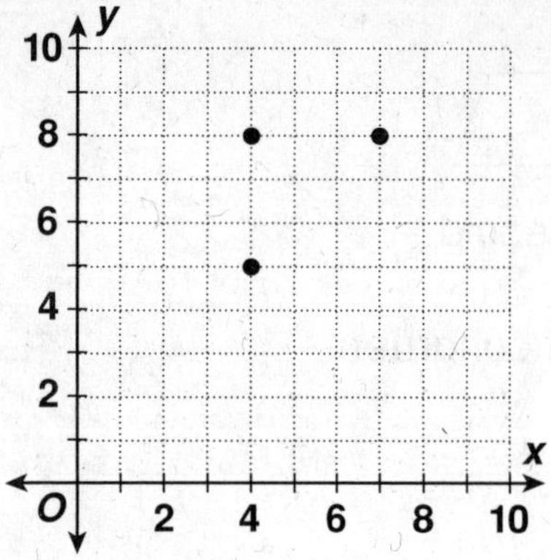

square

____

2.

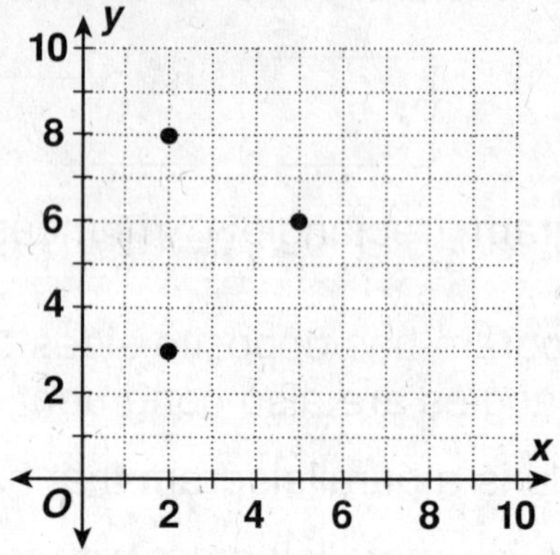

trapezoid

____

3.

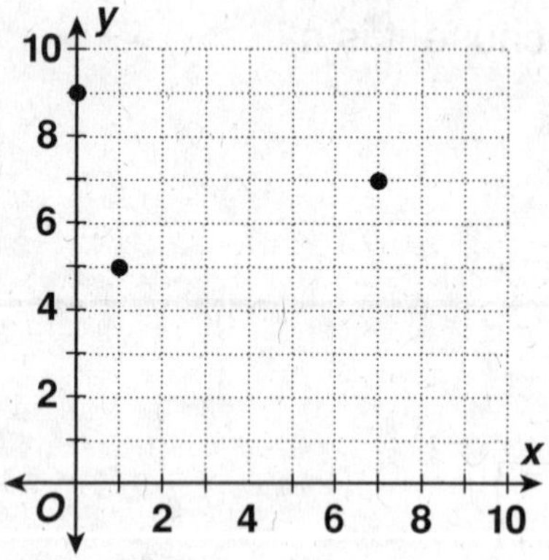

quadrilateral

____

4.

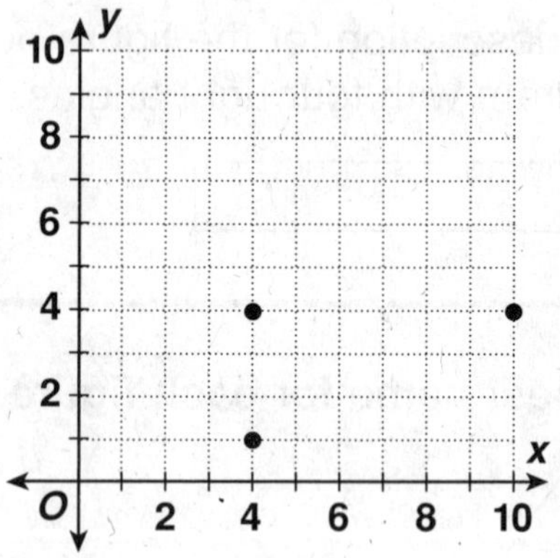

rectangle

____

5.

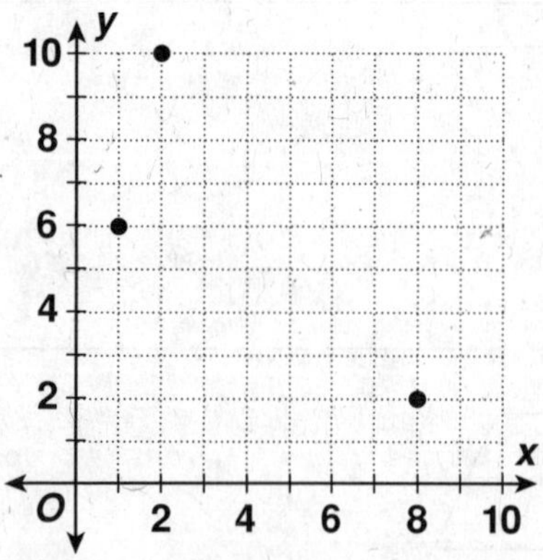

parallelogram

____

6.

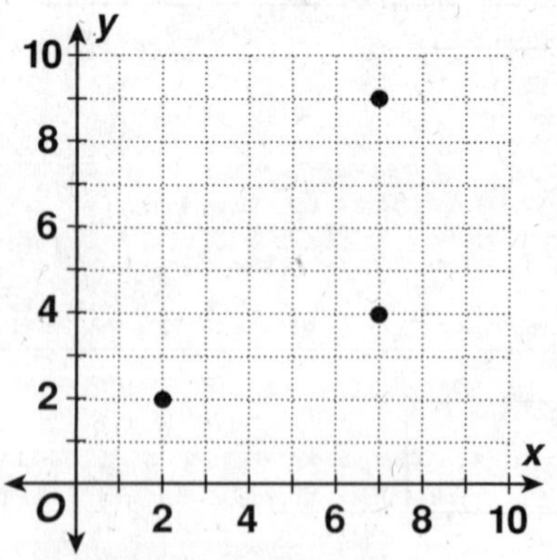

rhombus

____

Name ______________________ Date __________ Class __________

LESSON 8-6

Problem Solving

Quadrilaterals

Write the correct answer.

1. Fill in this Venn diagram using the terms quadrilaterals, squares, rectangles, rhombuses, parallelograms, and trapezoids.

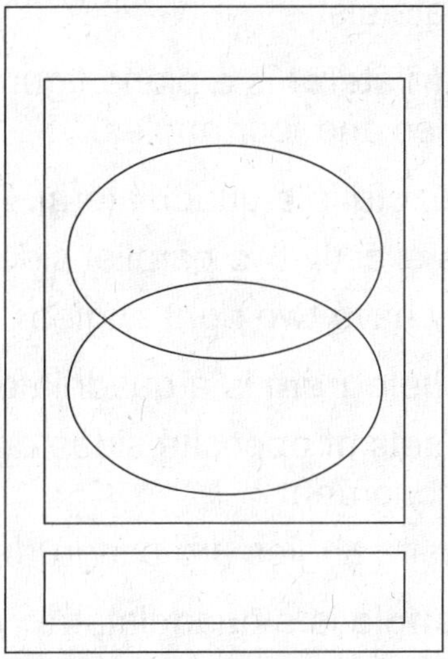

2. Part of this quadrilateral is hidden. What could it possibly be?

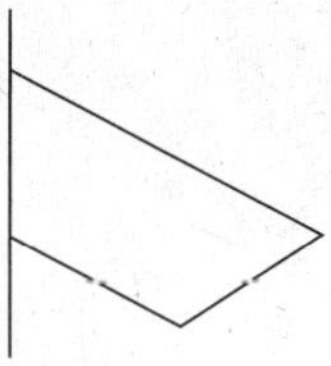

____

____

3. How could you make a trapezoid from a rectangle using only one cut?

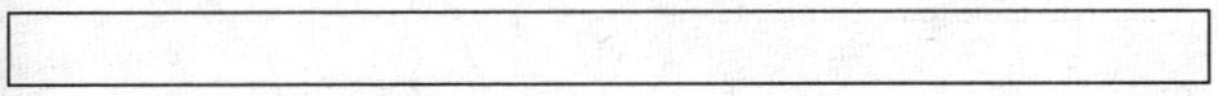

____

____

4. An engineer wants to build a building with a parallelogram base. He wants the four corners to be right angles and the four sides congruent. What type of base does the engineer want?

____

____

Circle the letter of the correct answer.

5. Each side of a quadrilateral-shaped picture frame has the same length. Which of the following is not a possible shape for the frame?

A a rhombus

B a square

C a trapezoid

D a parallelogram

6. The total length of the four sides of the picture frame from Exercise 5 is 4 feet, 8 inches. What is the length of each of its sides?

F 14 inches

G 1 foot, 3 inches

H 12 inches

J 2 inches

Name ______________________ Date ____________ Class ____________

LESSON 8-6

Reading Strategies

Focus on Vocabulary

The chart below will help you understand the different kinds of quadrilaterals.

A **quadrilateral** is a plane figure with four sides and four angles.

A **trapezoid** is a quadrilateral.

- It has exactly two parallel sides.
- It may have two right angles.

A **parallelogram** is a quadrilateral.

- Both sets of opposite sides are parallel and congruent.
- Opposite angles are congruent.

A **rectangle** is a quadrilateral.

- It is also a parallelogram with four right angles.

A **rhombus** is a quadrilateral.

- It is also a parallelogram with four congruent sides.

A **square** is a quadrilateral.

- It is also a parallelogram.
- It is also a rectangle with four congruent sides.
- It is also a rhombus with four right angles.

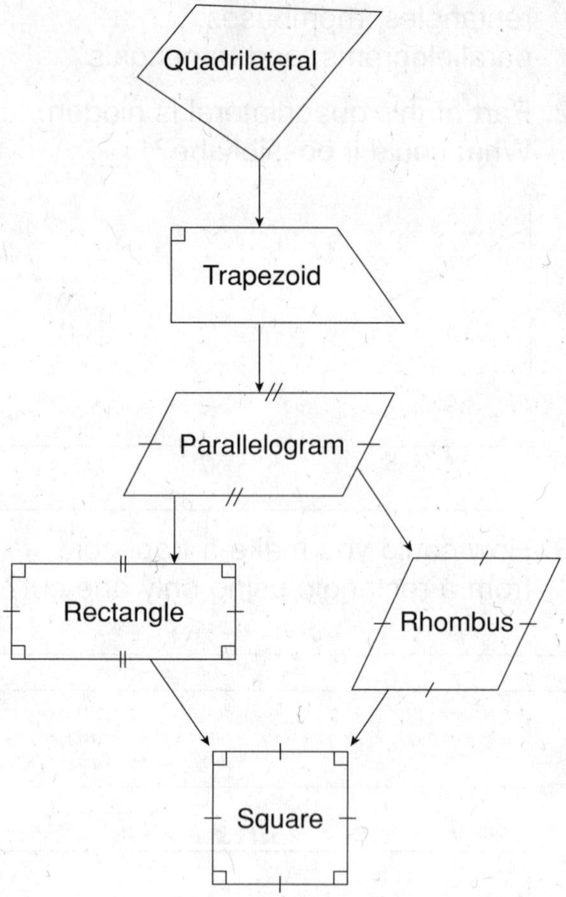

Use the chart to help you answer each question.

1. What is the name of a parallelogram with all sides equal? ____________

2. What is the name of a rectangle with all sides equal? ____________

3. What is the name of a quadrilateral with only one pair of parallel sides? ____________

4. Name the quadrilaterals that are also parallelograms.

____

Name ______________________ Date __________ Class __________

LESSON 8-6

Puzzles, Twisters & Teasers

Cafeteria Food

Decide if each statement is true or false, and circle the answer.

Answer the riddle by rearranging the letters next to your circled answers.

1. All squares are quadrilaterals.	True **S**	False **L**
2. A trapezoid is a quadrilateral with exactly two parallel sides.	True **R**	False **C**
3. Some squares are not rectangles.	True **O**	False **M**
4. A rhombus is a parallelogram with two congruent sides.	True **J**	False **O**
5. Some quadrilaterals are not squares.	True **M**	False **B**
6. Opposite angles of a parallelogram are not always congruent.	True **N**	False **U**
7. A rectangle is a parallelogram.	True **H**	False **K**
8. A triangle is a quadrilateral.	True **L**	False **O**

What do you call the place where school lunches are made? ______________

Name ______________________ Date ____________ Class ____________

LESSON **8-7**

Practice A
Polygons

Name each polygon.

1. 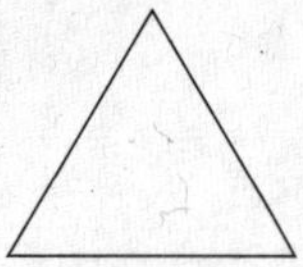

2. 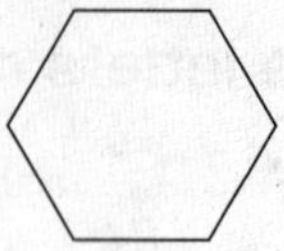

3. 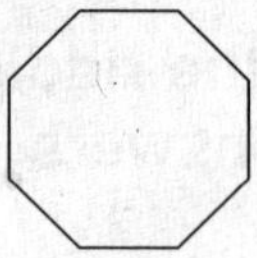

4. 

5. 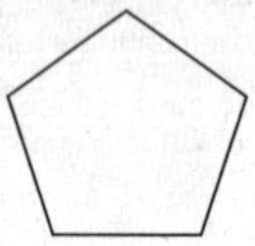

6. 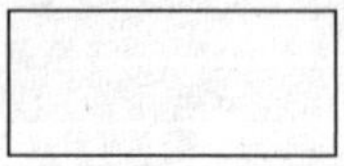

Tell whether each polygon appears to be *regular* or *not regular*.

7. 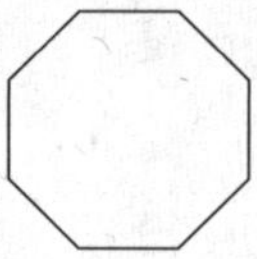

8. 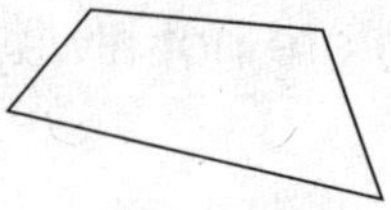

9. 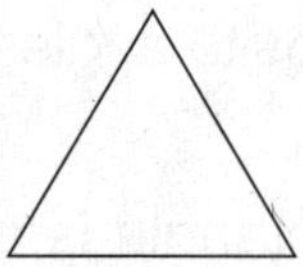

10. What is another name for a regular triangle?

11. What is another name for a regualr quadrilateral with four right angles?

12. One side of a regular pentagon measures 4 inches. What is the total length of the sides of that pentagon?

13. The total length of the sides of a regular octagon is 24 inches. What is the length of each side?

Name ________________ Date __________ Class __________

LESSON 8-7

Practice B

Polygons

Name each polygon and tell whether it appears to be *regular* or *not regular.*

1.

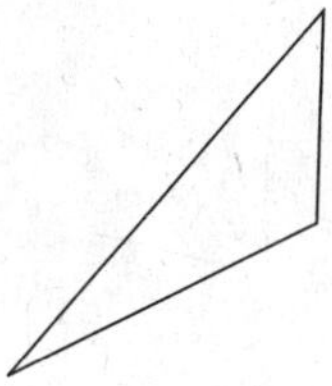

2.

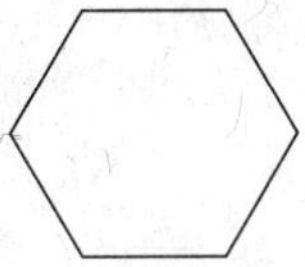

3.

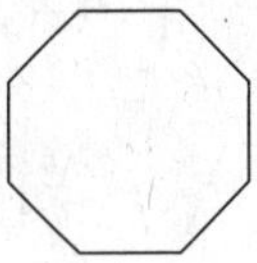

4.

5.

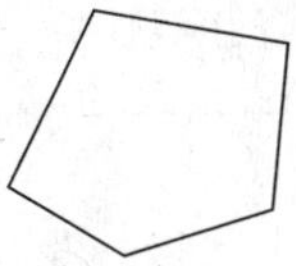

6.

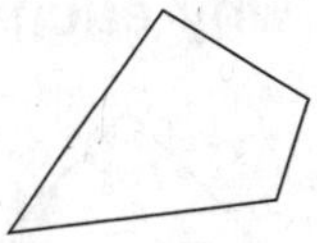

7. The public swimming pool is in the shape of a regular hexagon. Each side of the pool measures 5 feet. What is the distance around the entire pool?

8. In the space below, draw a regular quadrilateral. Now draw one diagonal of that quadrilateral. Describe the two polygons that are formed.

Name ______________________ Date ____________ Class ____________

LESSON **8-7**

Practice C
Polygons

Name each polygon and tell whether it appears to be *regular* or *not regular.*

1. 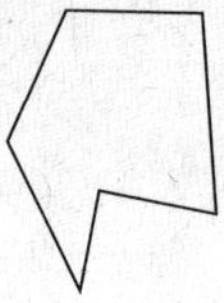

2. 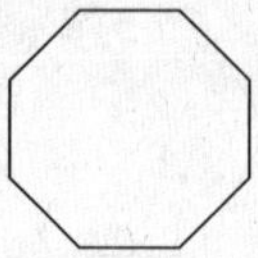

3. 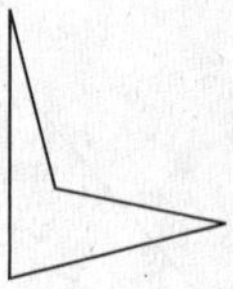

Explain why each shape is *not* a polygon.

4. 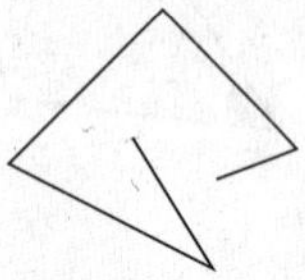

5. 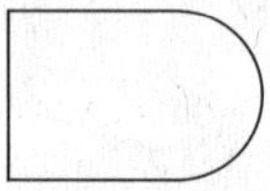

6. 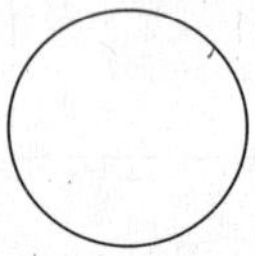

7. The perimeter of a flower bed designed in the shape of a regular hexagon is $28\frac{1}{2}$ feet. What is the length of each side of the flower bed?

8. How many diagonals can you draw from one vertex of a regular hexagon? What is the least number of triangles needed to fill a regular hexagon?

9. What is the total number of diagonals that can be drawn in a regular pentagon? Describe the figure formed by the diagonals.

Name ______________________ Date ____________ Class ____________

LESSON 8-7

Reteach

Polygons

A polygon is a closed plane figure formed by three or more line segments.

Polygons are named by the number of their sides and angles.

Sides and Angles	**3**	**4**	**5**	**6**	**8**
Polygon	triangle	quadrilateral	pentagon	hexagon	octagon

A regular polygon has all congruent sides and angles.

Each angle is 90 degrees.
Each side is 3 centimeters long.

If the angles and the sides are not all congruent, then the polygon is not regular.

Name each polygon and tell whether it is regular or not regular.

1.

____

2.

____

To find the angle measures of a regular hexagon, first divide the polygon into triangles by drawing connecting segments from one vertex to all of its nonadjacent vertices.

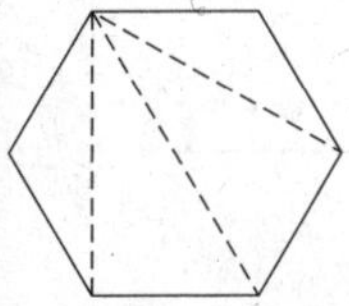

There are 4 triangles.

Next, multiply the number of triangles by 180.

$4 \cdot 180 = 720$ The sum of the angle measures of a triangle is 180°.

Divide the product by 6, the number of sides, to find the measure of each angle.

$\frac{720}{6} = 120$ All of the angles of a regular polygon are congruent.

Each angle is 120°.

Find the angle measure of each regular polygon.

3. regular quadrilateral ____________

4. regular pentagon ____________

Name ______________________ Date ____________ Class ____________

LESSON 8-7

Challenge

Sign Language

Traffic signs are symbols used to convey information. In the United States, certain shapes are used only for certain traffic signs. As a result, you can often understand the message of a traffic sign simply by classifying its geometric shape.

Tell whether each traffic sign below is a polygon. If it is, name the polygon that best describes its shape, and tell whether it appears to be *regular* or *not regular*.

1.

2.

3.

4.

5.

6.

7.

8.

9.

Name ______________________ Date ____________ Class ____________

LESSON **8-7**

Problem Solving
Polygons

Write the correct answer.

1. Name each polygon in this figure.

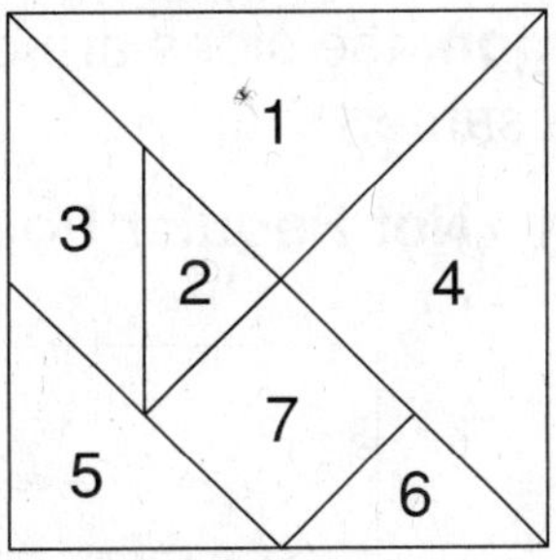

____

____

____

____

2. How could you use the sum of the angles inside a triangle to find the sum of the angles inside a heptagon?

____

____

____

____

3. How could you use the sum of the angles inside a triangle to find the sum of the angles inside a decagon?

____

____

____

____

4. In the space below, draw a rectangle and a parallelogram with side lengths congruent to the rectangle's. Now draw the diagonals for each of those polygons. What new polygons are formed by the diagonals in each quadrilateral?

____

5. In Exercise 4, what is true of the diagonals in the rectangle that isn't true of the diagonals of the parallelogram?

____

____

Circle the letter of the correct answer.

6. The perimeter of a regular hexagon is $13\frac{1}{2}$ inches. What is the length of each side?

A $2\frac{7}{10}$ inches

B $2\frac{1}{4}$ inches

C $3\frac{3}{4}$ inches

D $1\frac{11}{16}$ inches

7. Which of the following statements is sometimes false?

F A plane figure is a polygon.

G Each side of a polygon intersects exactly two other sides.

H A polygon is a closed figure.

J A polygon has straight sides.

Name ______________________ Date __________ Class __________

LESSON **8-7**

Reading Strategies
Focus on Vocabulary

A **polygon** is a closed, plane figure formed by three or more line segments.

In a **regular polygon,** the sides are of equal length and all angle measures are the same.

Regular Polygon **Not Regular Polygon**

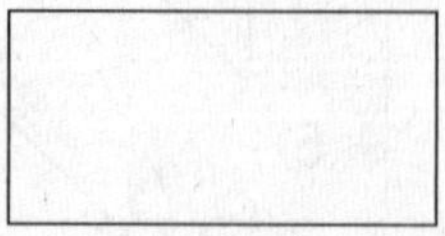

The prefix of a polygon is a clue that can help you understand the number of sides and angles it has.

- *Tri-* means "three" → A **tri**ple play means three outs all at once.
  → A **tri**angle has three sides and three angles.
- *Quad-* means "four" → **Quad**ruplets are four babies born at the same time.
  → A **quad**rilateral has four sides and four angles.

This table helps you organize polygons by their sides and angles.

Polygon	Number of Sides	Number of Angles
Triangle	3 sides	3 angles
Quadrilateral	4 sides	4 angles
Pentagon	5 sides	5 angles
Hexagon	6 sides	6 angles
Octagon	8 sides	8 angles

Write the name of each polygon. Also write whether the polygon is *regular* or *not regular*.

1.

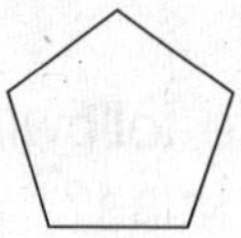

____

2.

____

3.

____

4.

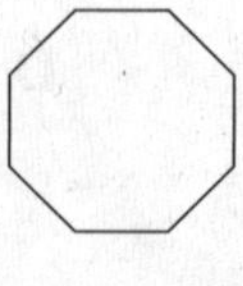

____

Name ______________________________ Date ____________ Class ____________

LESSON **8-7**

Puzzles, Twisters & Teasers

Seeing is Believing

There are three correct answers for each problem. Circle the single incorrect statement. The letters next to the incorrect statement will provide the solution to the riddle when they are placed on the blank lines, which match the numbers below.

1.

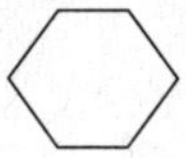

T has 6 angles
F is an octagon
N is an irregular polygon
A has 6 sides

2.

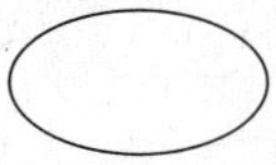

U has no sides
E is an oval
A is not a quadrilateral
Y is a polygon

3.

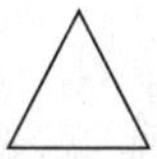

L has 3 sides
K is an isosceles triangle
P is not a polygon
D sum of interior angles equals 180°

4.

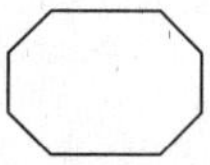

O is an octagon
A sum of the interior angles is 720°
G can be divided into 6 triangles
S has 8 angles

5.

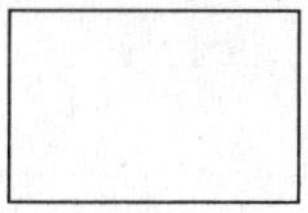

P has 4 right angles
L is not a quadrilateral
H has 4 sides
W sum of interior angles equals 360°

6.

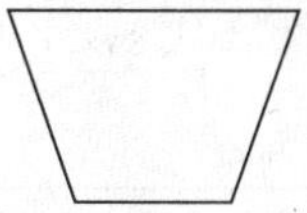

K is a polygon
S has 4 congruent sides
N is a trapezoid
T is a quadrilateral

7.

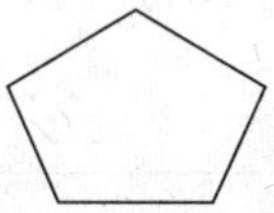

O sum of interior angles equal 560°
U has 5 sides
E can be divided into 3 triangles
A has 5 angles

8.

J is a square
M is a regular polygon
F sum of the interior angles is 360°
C sides are not congruent

What kind of glasses do spies use?

___	___	___	___	___	___	___	___	___
6	3	2	1	7	8	4	5	6

Name ______________________ Date __________ Class __________

LESSON 8-8

Practice A

Geometric Patterns

Identify a possible pattern. Use the pattern to draw the next figure.

1.

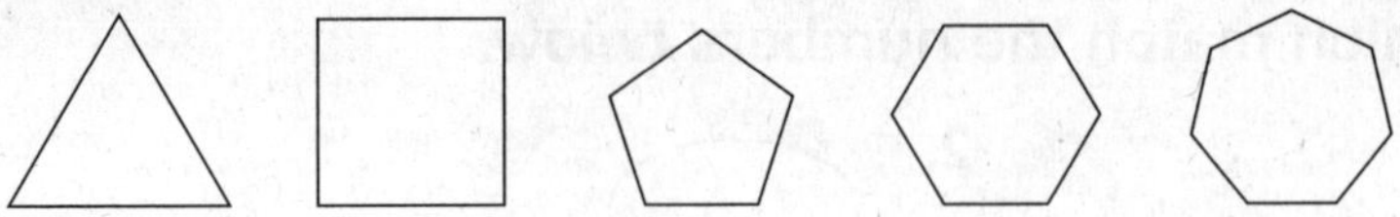

2.

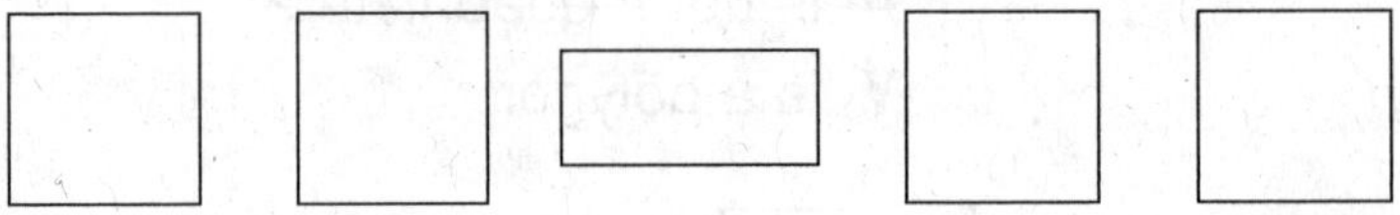

3.

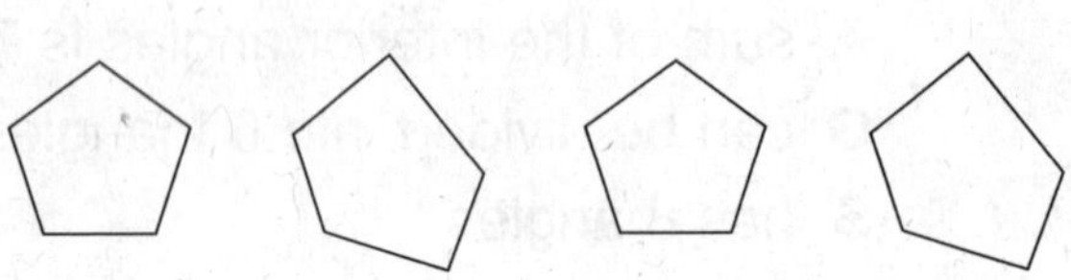

4.

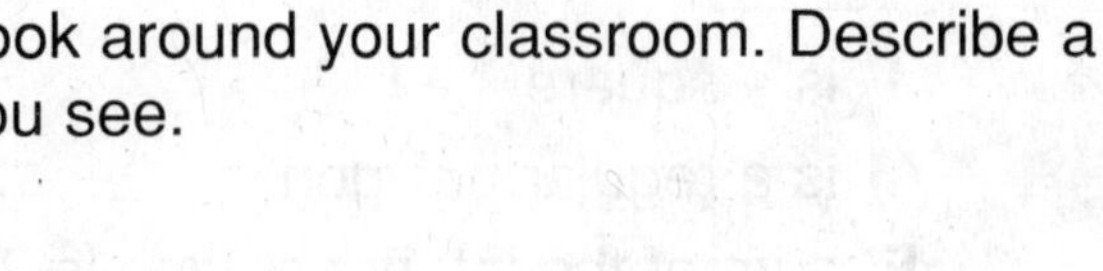

5. Look around your classroom. Describe a geometric pattern you see.

6. Use squares to create a geometric pattern. Describe your pattern.

Name ______________________ Date ____________ Class ____________

LESSON **8-8**

Practice B

Geometric Patterns

Identify a possible pattern. Use the pattern to draw the next figure.

1.

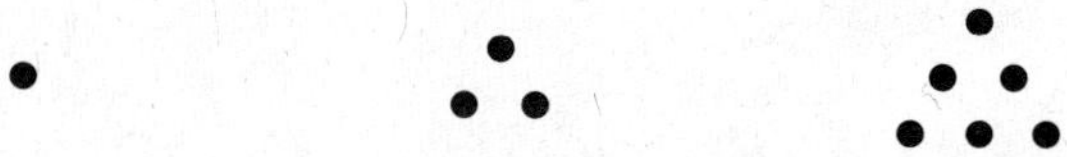

____

____

2.

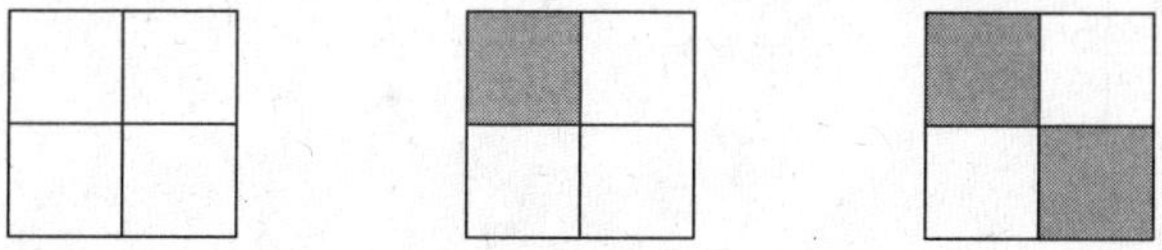

____

3.

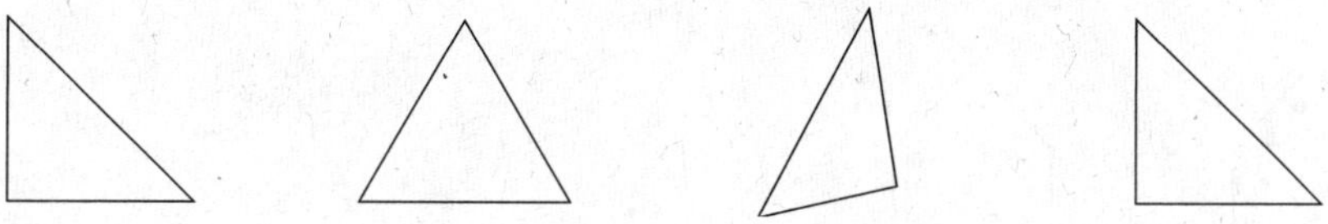

____

____

4.

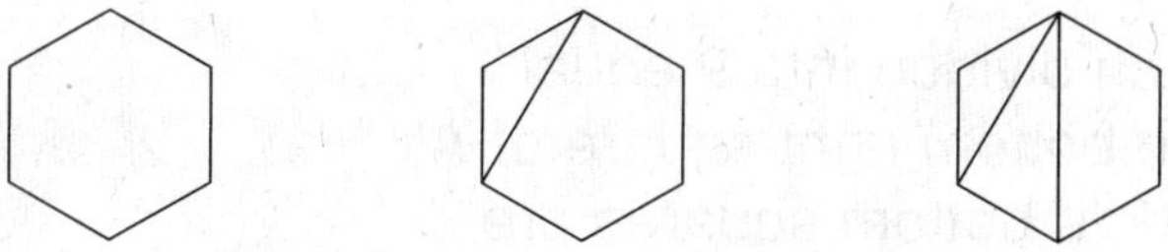

____

____

5. Use triangles to create a geometric pattern. Describe your pattern.

____

____

Name ______________________ Date __________ Class __________

LESSON 8-8

Practice C

Geometric Patterns

Identify a possible pattern. Use the pattern to draw the next figure.

1.

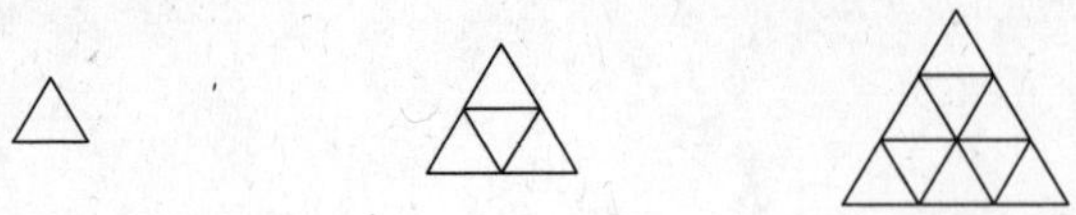

____

____

2.

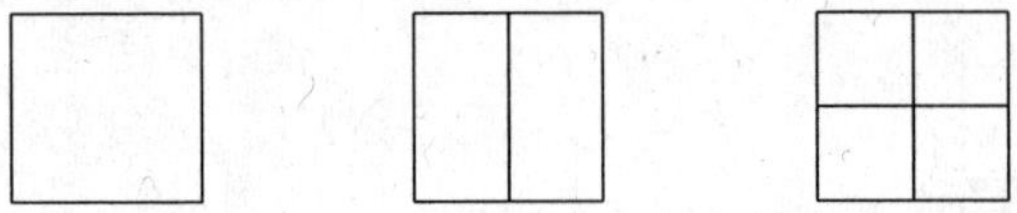

____

____

3.

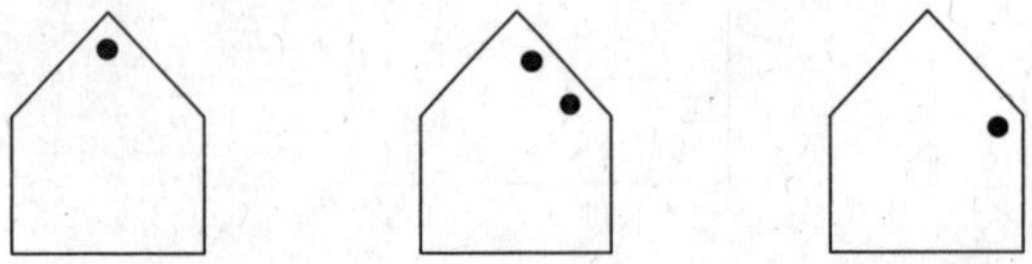

____

____

4. Three squares, **A**, **B**, and **C**, are each divided into 9 equal squares. The top left square and the bottom right square of **A** are shaded. The middle top and middle bottom squares are shaded in **B**. In **C**, the top right and the bottom left squares are shaded. Describe **D**, the next square in the pattern.

____

____

5. Use pentagons to create a geometric pattern. Describe your pattern.

____

____

Name ______________________ Date ____________ Class ____________

LESSON 8-8

Reteach

Geometric Patterns

Sometimes, patterns involve geometric shapes and figures. The pattern can be based on shape, color, size, position, or quantity.

Before you draw a missing figure in a pattern, you first have to look at the other figures to recognize a relationship among the figures.

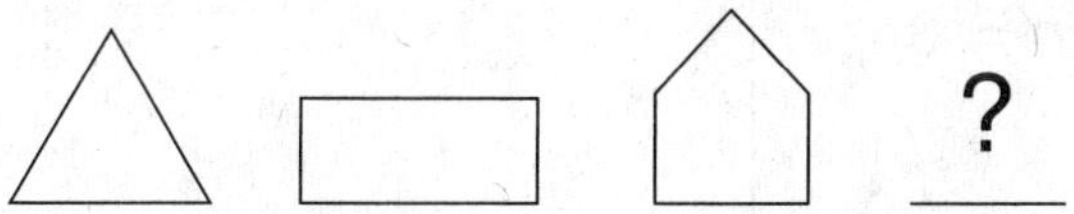

In the pattern above, each figure has one more side than the figure that proceeds it.

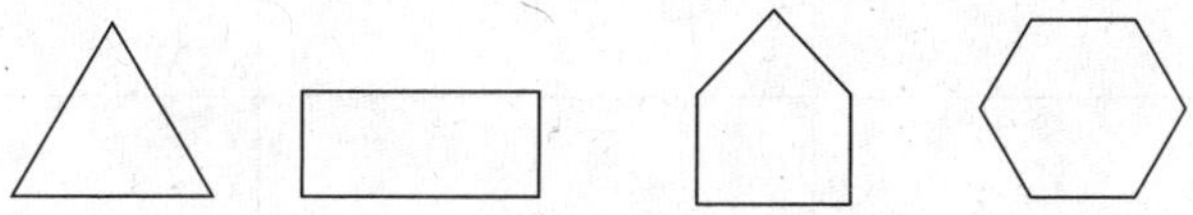

To continue the pattern, the next figure would be a hexagon.

Identify a pattern for each. Then draw the missing figure.

1.

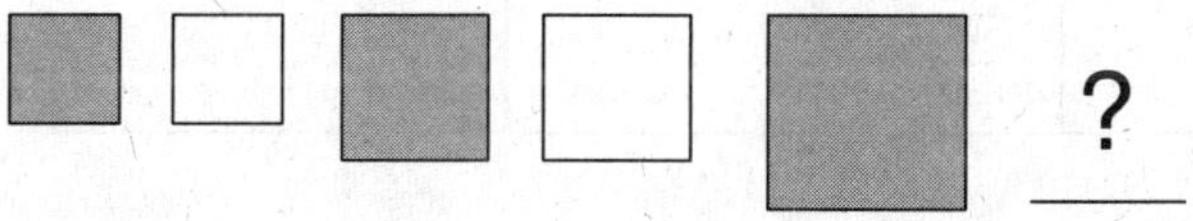

2.

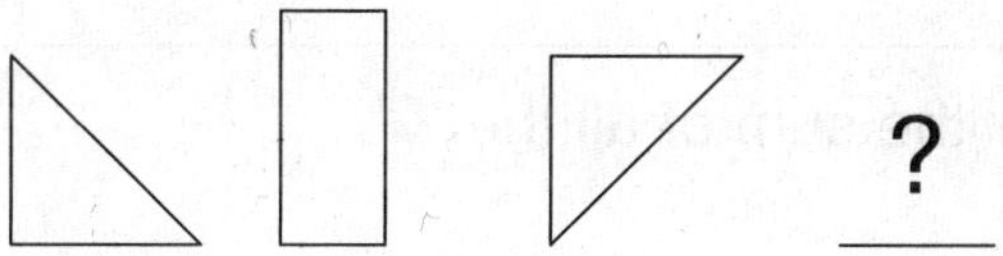

3.

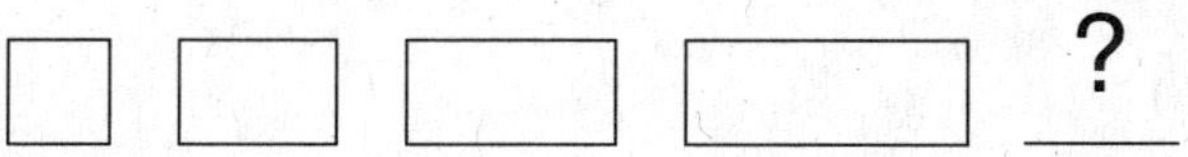

Name ______________________ Date __________ Class __________

LESSON **8-8**

Challenge

Polygon Patterns

Look for patterns to complete this chart and discover two rules for polygons.

Regular Polygon	Number of Sides, *n*	Number of Triangles to Fill	Sum of the Interior Angles
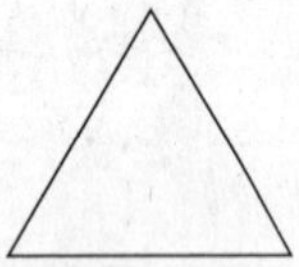 Triangle	$n = 3$	1	$1(180°) = 180°$
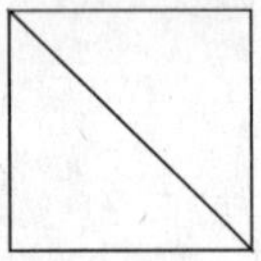 Quadrilateral			$2(180°) = 360°$
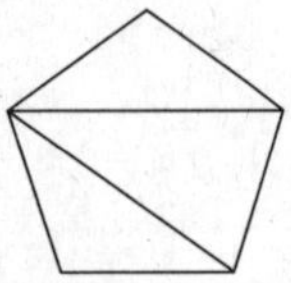 Pentagon			
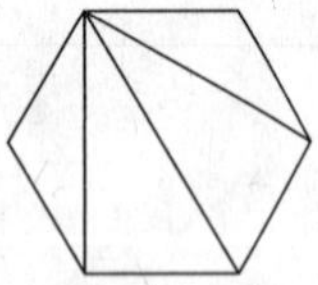 Hexagon			

1. Use the patterns to write an expression for the number of triangles needed to fill any regular polygon.

____

2. Use the patterns to write an expression for the sum of all the angles inside any regular polygon.

____

3. Using these expressions, how many triangles are needed to fill any regular octagon? What is the sum of a regular octagon's interior angles?

____

Name ______________________ Date __________ Class __________

LESSON 8-8 Problem Solving

Geometric Patterns

Complete this chart and look for patterns. Then answer the questions.

	Number of Points on the Line	**Draw and Label the Line and Points**	**Number of Different Line Segments in the Line**
	2	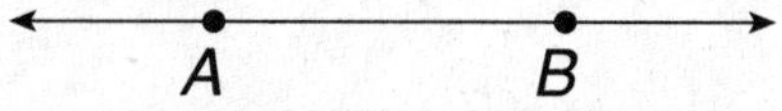	1
1.		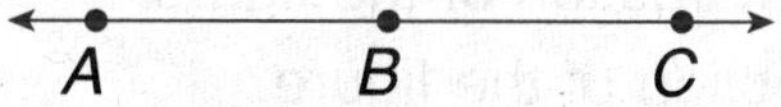	
2.		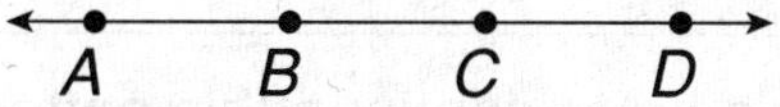	
3.		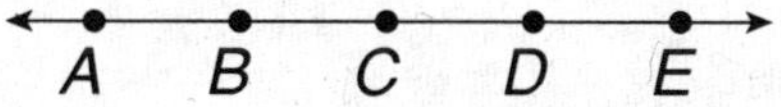	
4.		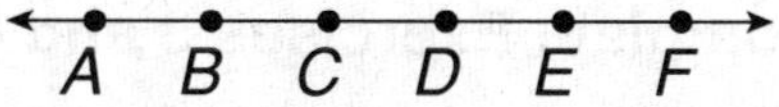	

Circle the letter of the correct answer.

5. If n = the number of points on a line, which of the following expressions shows the number of different line segments on that line?

A $2n - 3$

B $(n^2 - n) \div 2$

C $(n \div 2) \cdot 5$

D $10n \div 2$

6. Using the pattern in the table and your answer to Exercise 5, how many different line segments will be on a line if there are 10 points on the line?

F 17 line segments

G 25 line segments

H 45 line segments

J 50 line segments

Name ______________________ Date ____________ Class ____________

LESSON 8-8

Reading Strategies

Analyze Information

You can examine geometric figures and notice a pattern that forms from one figure to the next.

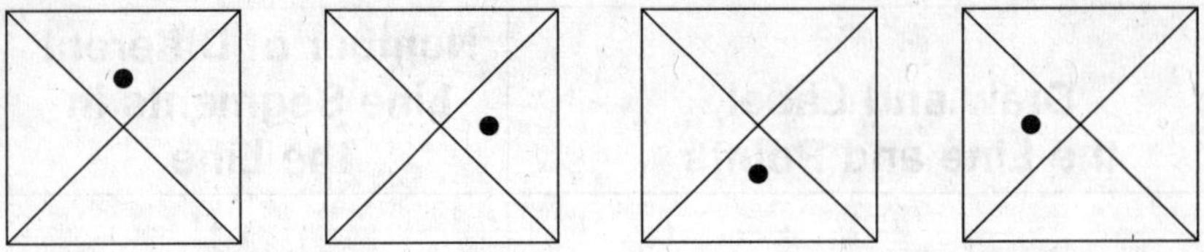

- The black dot is in the **top** section of the figure.
- The black dot is in the **right** section of the figure.
- The black dot is in the **bottom** section of the figure.
- The black dot is in the **left** section of the figure.

1. If the pattern continues, where will the dot be in the next figure?

____

2. Draw the dot where it should be to continue this pattern.

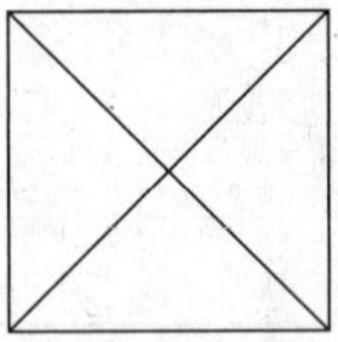

Examine these figures and complete Exercises 3 and 4.

3. What do you notice about the number of dots in the bottom row of each figure?

____

____

4. Draw the next figure to continue the pattern.

Name ________________________ Date ____________ Class ____________

LESSON 8-8

Puzzles, Twisters & Teasers

What's Next?

Draw lines to match the next pattern in the sequence. Not all patterns will match. Once you have all the lines drawn go back to number one and put the letter that is next to the answer in the puzzle answer.

1.	
2.	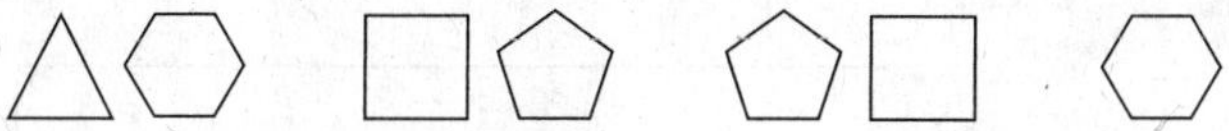
3.	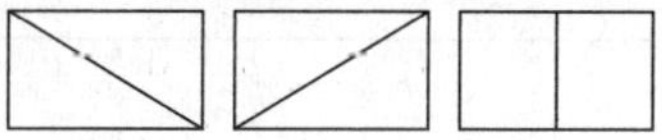
4.	
5.	
6.	
7.	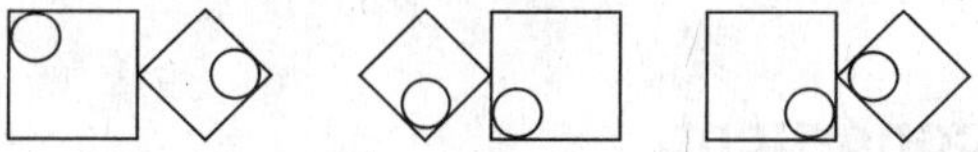

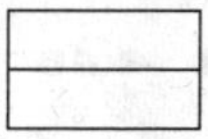	**K**
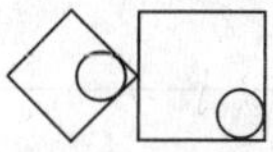	**T**
	**L**
	**A**
	**S**
	**I**
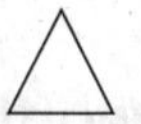	**P**
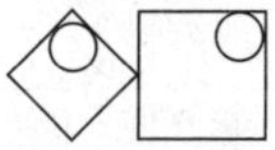	**E**
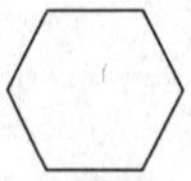	**L**
	**C**

What do you call a cucumber in a sour mood?

____   ____ ____ ____ ____ ____ ____

1   2 5 6 3 4 7

Name ______________________ Date ____________ Class ____________

LESSON 8-9

Practice A

Congruence

Decide whether the figures in each pair are congruent. If not, explain.

1.

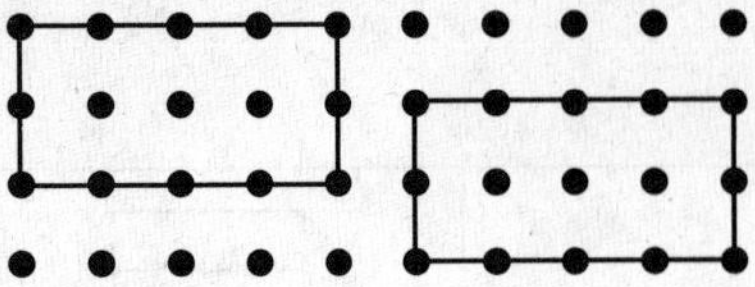

____

____

2.

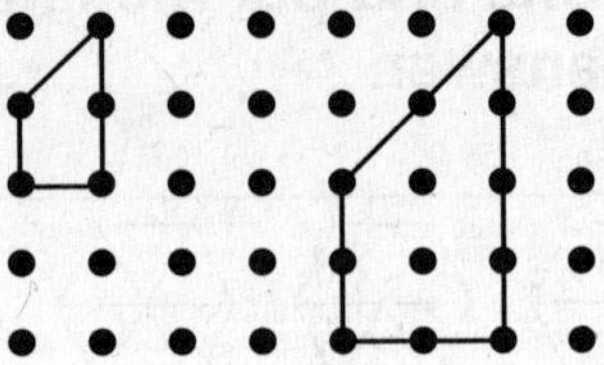

____

____

3.

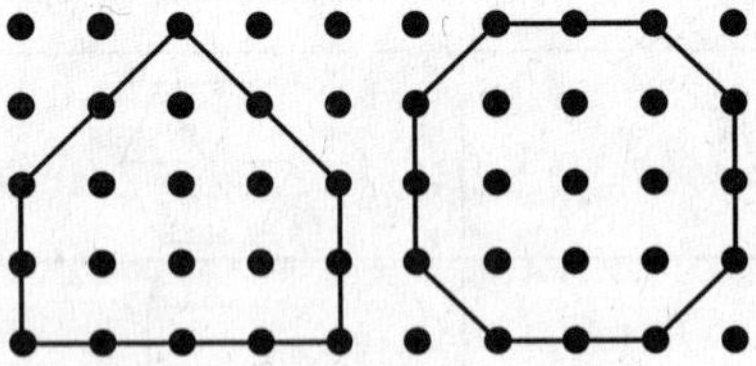

____

____

4.

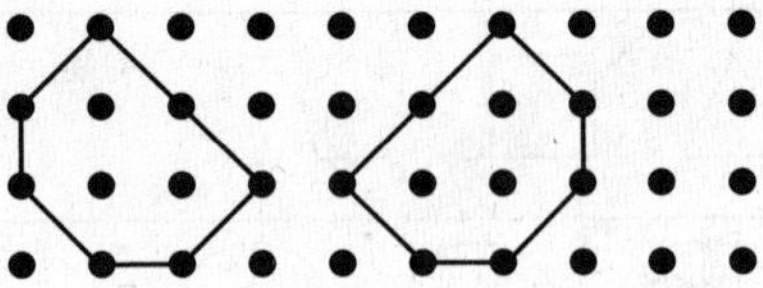

____

____

Circle the two figures in each row that are congruent.

5.

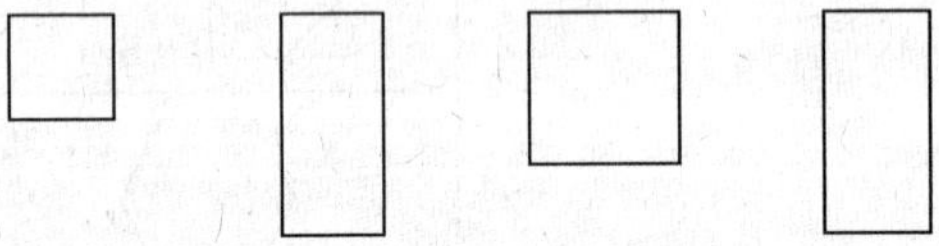

6.

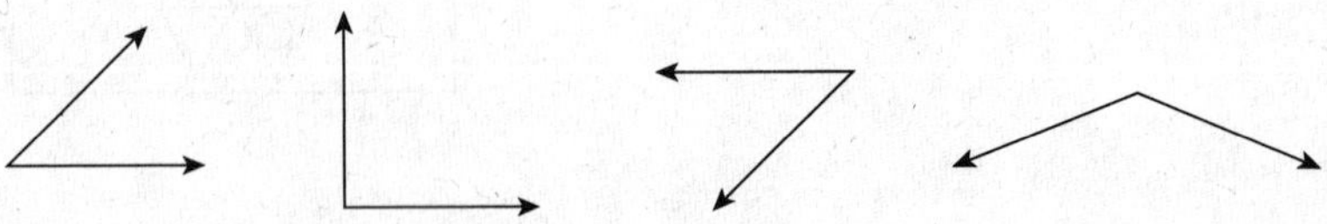

7. Use a ruler to draw two congruent line segments.

____

8. If you know that two figures are congruent, what do you know about them?

____

Name ______________________ Date __________ Class __________

LESSON 8-9

Practice B
Congruence

Decide whether the figures in each pair are congruent. If not, explain.

1.

____

____

2.

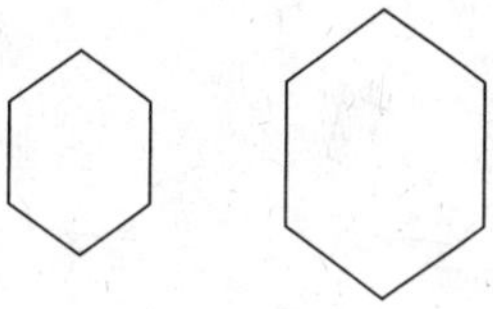

____

____

3.

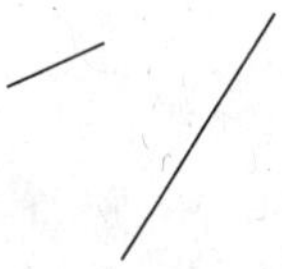

____

____

4.

____

____

Use the diagram for Exercises 5–7.

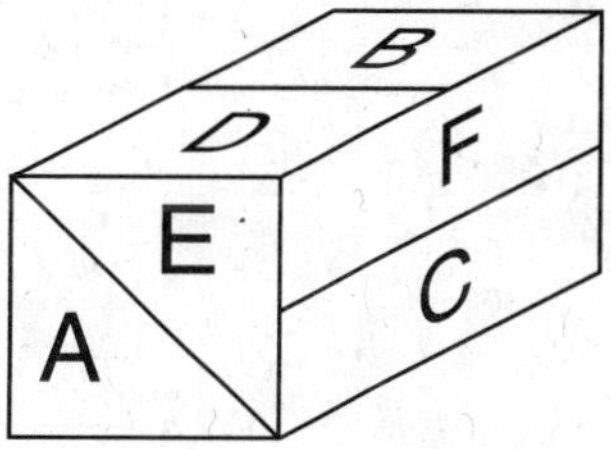

5. Which part of the figure is congruent to A? __________

6. Which part of the figure is congruent to D? __________

7. Which part of the figure is congruent to F? __________

8. Name two parts of your body that appear to be congruent.

____

9. Square *ABCD* is congruent to square *FGHJ*. The total length of the sides of square *ABCD* is 12 meters. What is the length of each side of square *FGHJ*?

____

Name ______________________ Date __________ Class __________

LESSON 8-9

Practice C
Congruence

Decide whether the figures in each pair are congruent. If not, explain.

1.

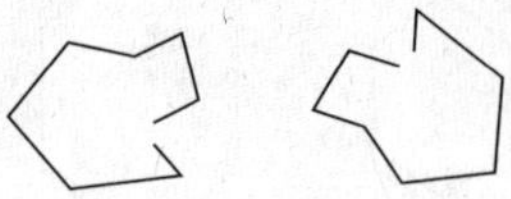

____

____

2.

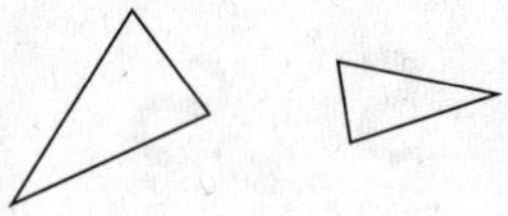

____

____

Use the dot paper to draw a congruent figure next to the given figure.

3.

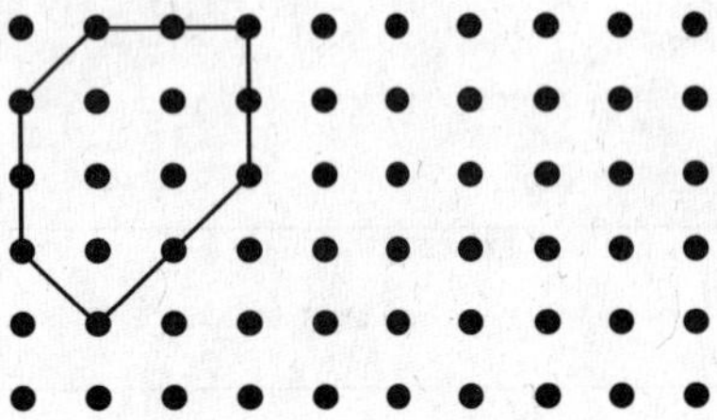

4.

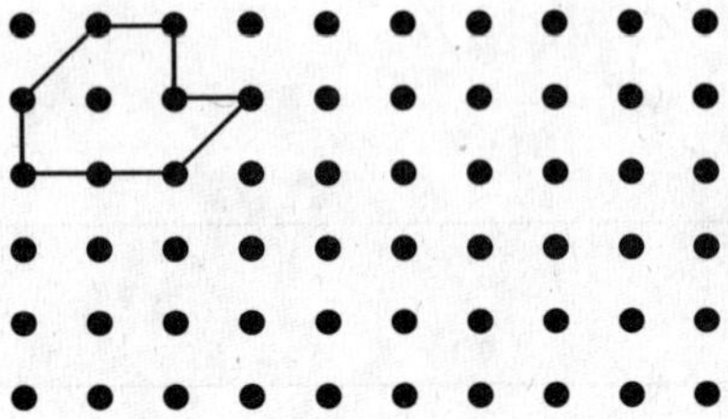

5. Draw a line inside this triangle to create two congruent figures. Describe what type of line you drew and classify the congruent figures you created.

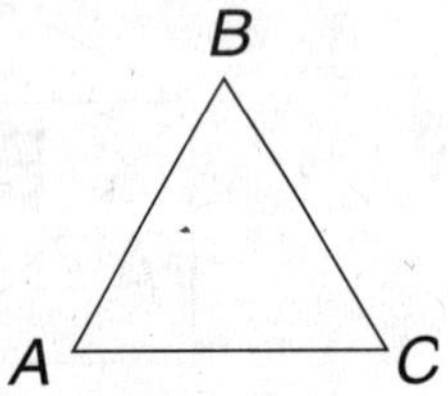

____

____

6. Name all the congruent parts of this glass box.

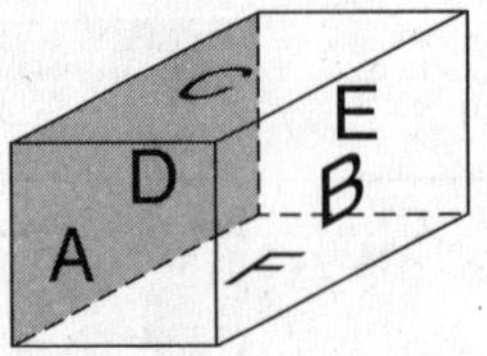

____

Name ________________________ Date ____________ Class ____________

LESSON **8-9**

Reteach

Congruence

Two figures are congruent if they have the same size and shape.

Look at the two figures.

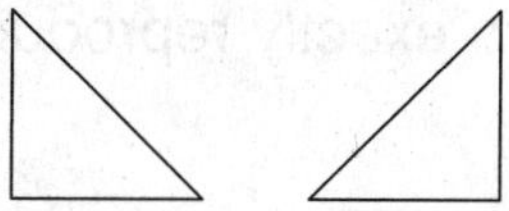

They are congruent because they are both triangles and they are the same size.

Now look at these two figures.

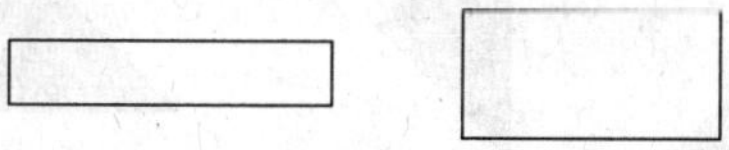

These figures are not congruent. They are both rectangles but they are not the same size.

Decide whether the figures in each pair are congruent. If not, explain.

1.

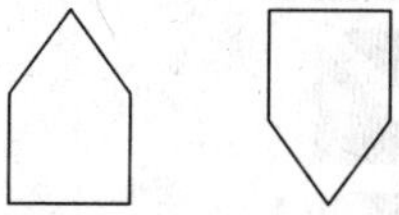

____

2.

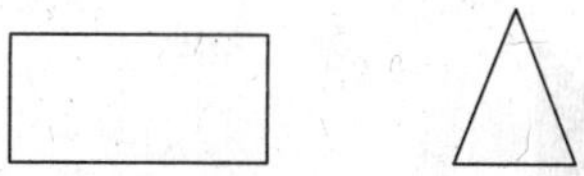

____

3.

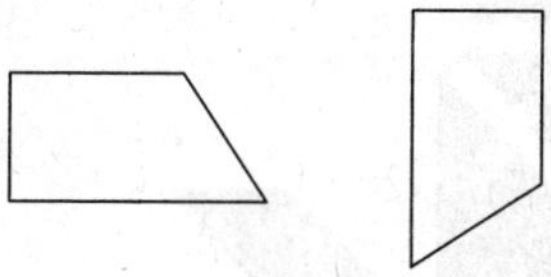

____

Name ______________________ Date ____________ Class ____________

LESSON 8-9

Challenge

Tangram

Tangram is an ancient Chinese puzzle made by cutting a square into 7 pieces—2 congruent small triangles, 2 congruent large triangles, 1 medium triangle, a square, and a parallelogram. To solve a Tangram, you need to rearrange those seven pieces to exactly reproduce a given image.

The seven Tangram puzzle pieces are shown below. Trace the pieces and cut them out. Then use them to solve each Tangram that follows.

1.

2.

Name ______________________ Date ____________ Class ____________

LESSON 8-9

Problem Solving

Congruence

Write the correct answer.

1. Similar figures have the same shape but may have different sizes. How are similar figures different from congruent figures?

2. Pentagon A and Pentagon B are congruent regular polygons. If the total length of the sides of Pentagon B is 68.5 feet, what is the length of each side of Pentagon A?

3. Is the following statement always true, sometimes true, or never true? Two congruent figures are similar figures. Explain.

4. Draw a figure congruent to this line segment. Explain how you drew your congruent figure.

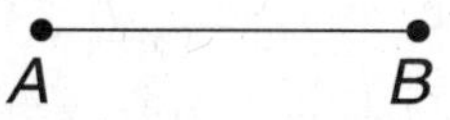

Circle the letter of the correct answer.

5. Which word makes this statement true? Corresponding parts of congruent figures are ________.

A not regular
B congruent
C polygons
D horizontal

6. If two angles of a right triangle are congruent, what are the measures of each angle in the triangle?

F 35°, 55°, and 90°
G 45°, 45°, and 90°
H 50°, 50°, and 90°
J 55°, 55°, and 90°

7. Which of the following polygons do not always have all congruent sides?

A a square
B an equilateral triangle
C a rhombus
D a pentagon

8. If $\angle A$ of rectangle *ABCD* is congruent to $\angle X$ of triangle *XYZ*, which of these statements is true?

F Rectangle *ABCD* is also a square.
G Triangle *XYZ* is a right triangle.
H Rectangle *ABCD* is a regular polygon.
J Triangle *XYZ* is an acute triangle.

Name ______________________ Date ____________ Class ____________

LESSON 8-9

Reading Strategies
Graphic Organizer

This chart helps you understand **congruence.**

Definition

Two figures that have exactly the same size and the same shape are congruent.

Facts

- Figures have same size
- Figures have same shape

Congruence

Examples

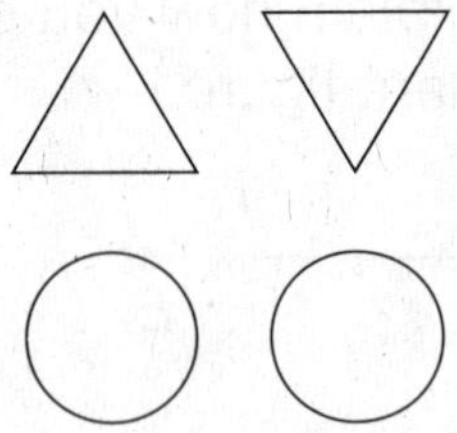

Non-examples

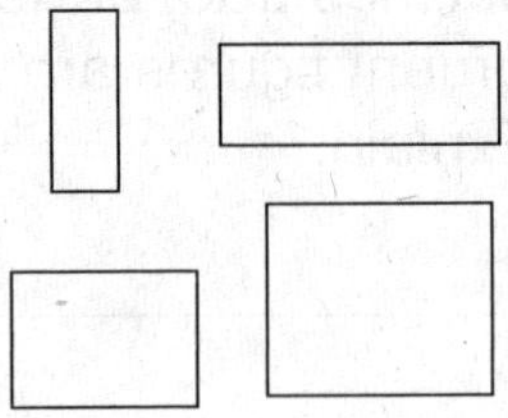

Use the chart to answer each question.

1. How do you know the triangles are congruent?

____

____

2. Why aren't the rectangles congruent?

____

____

Write *congruent* or *not congruent* to describe each pair of figures.

3.

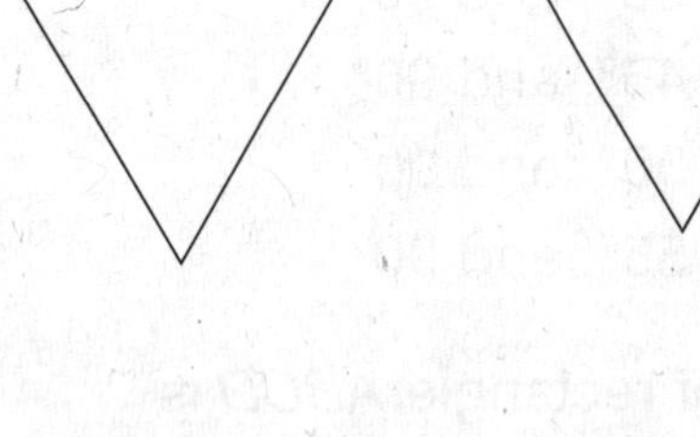

____

4.

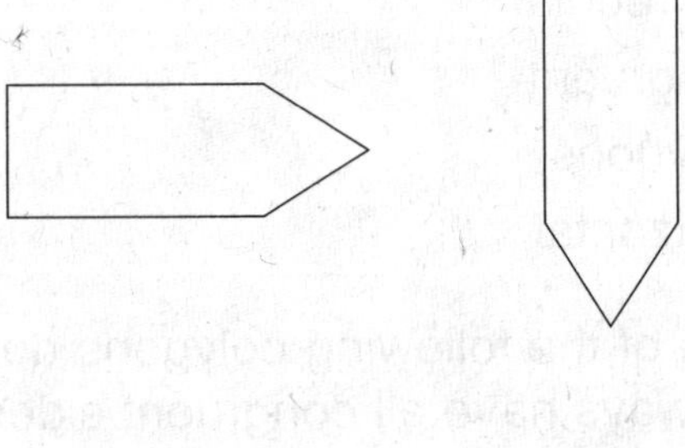

____

5.

____

6.

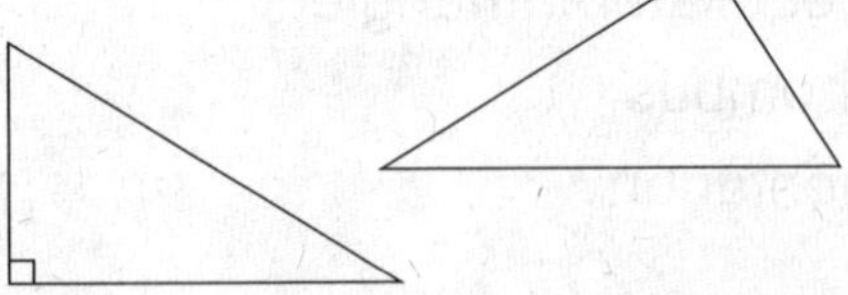

____

Name ______________________________ Date ____________ Class ____________

LESSON **8-9**

Puzzles, Twisters & Teasers

What is the difference?

Identify the pair of congruent figures in each row. Notice the letter next to each correct answer. Write the letter on the blank line which corresponds to the problem you have just solved. (There is only one correct answer per problem.)

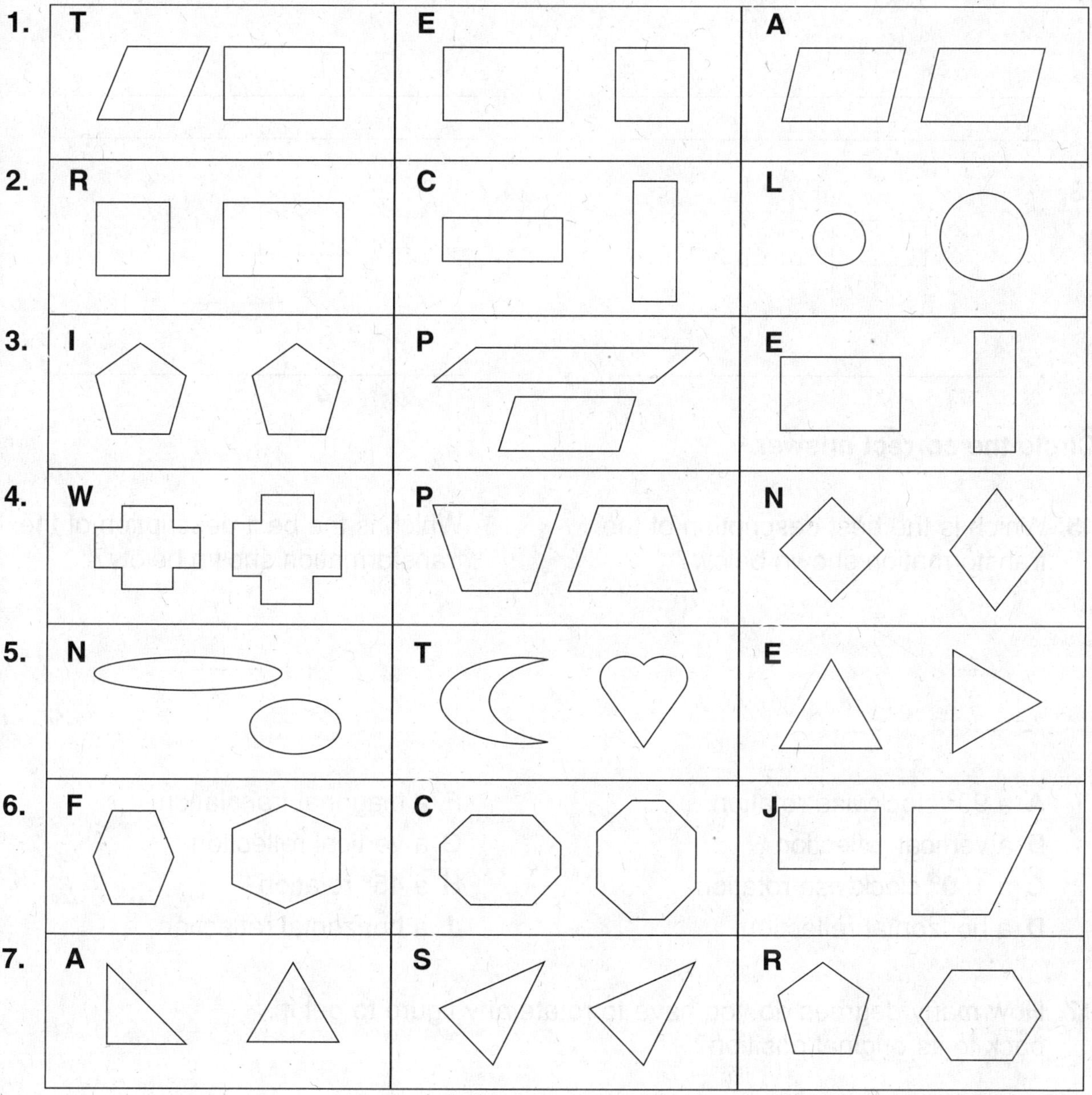

What do polar bears wear on their heads?

___ ___ ___   ___ ___ ___ ___
3 2 5   6 1 4 7

Name ______________________ Date ____________ Class ____________

LESSON **8-10**

Practice A

Transformations

Tell whether each is a translation, rotation, or reflection.

1.

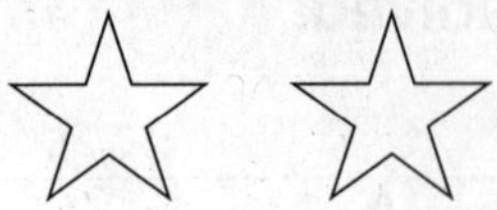

____

____

2.

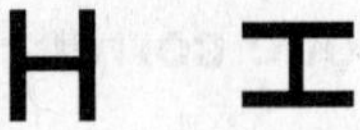

____

____

3.

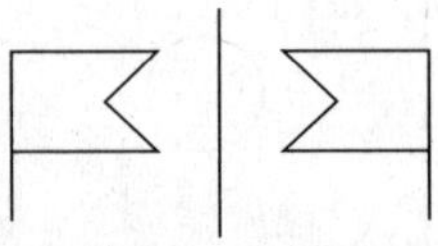

____

4.

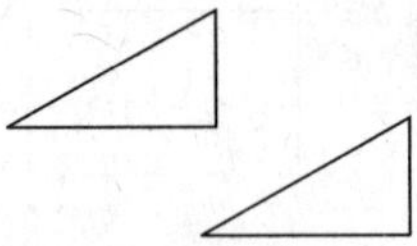

____

Circle the correct answer.

5. Which is the best description of the transformation shown below?

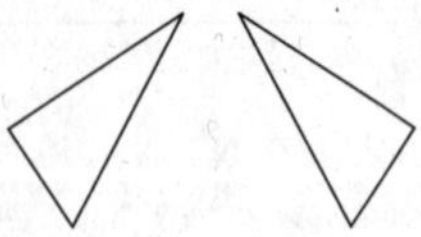

A a 90° clockwise rotation
B a vertical reflection
C a 180° clockwise rotation
D a horizontal reflection

6. Which is the best description of the transformation shown below?

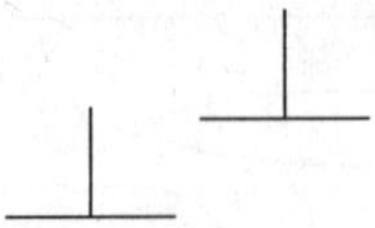

F a diagonal translation
G a vertical reflection
H a 45° rotation
J a horizontal reflection

7. How many degrees do you have to rotate any figure to get it back to its original position?

____

8. Why do words look different when you hold them up to a mirror?

____

Name ______________________ Date ____________ Class ____________

LESSON **8-10**

Practice B

Transformations

Tell whether each is a translation, rotation, or reflection.

1.

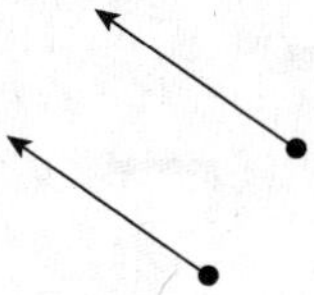

____

2.

____

3.

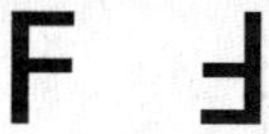

____

Draw each transformation.

4. Draw a 180° clockwise rotation about the point.

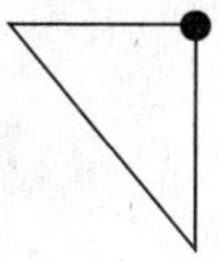

5. Draw a vertical reflection across the dotted line.

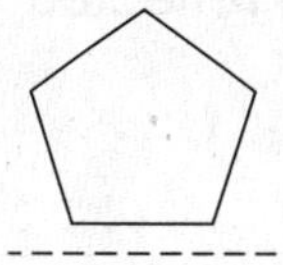

6. Without using reflections, how can you get this △ to look like this ▽?

____

____

7. Describe a horizontal reflection of the word **MOM**. Can you think of any other words that would have a similar horizontal reflection?

____

____

Name ________________________ Date ____________ Class ____________

LESSON **8-10**

Practice C

Transformations

Tell whether each of the transformations of each figure is a translation, rotation, or reflection.

1.

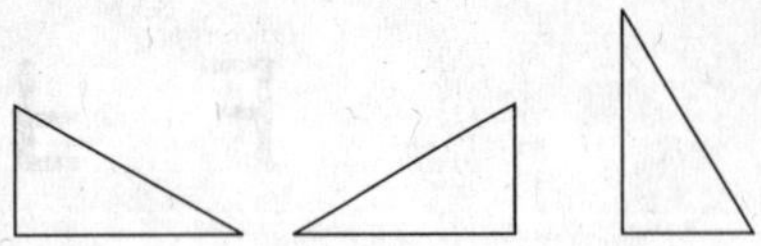

____

____

2.

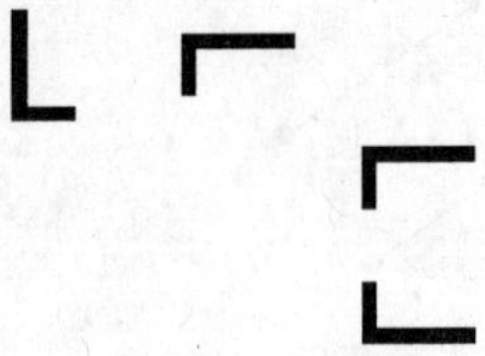

____

____

Draw each transformation.

3. Draw a horizontal reflection across the dotted line. Then draw a 90° clockwise rotation of the reflected figure.

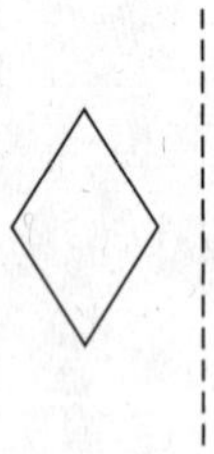

4. Draw a 180° counterclockwise rotation about the point. Then translate the rotated figure.

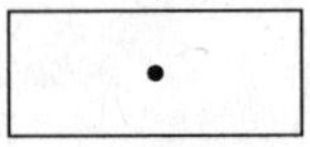

5. Name five capital letters that look the same before and after a reflection.

____

6. Explain how folding a piece of paper could help you check your reflections.

____

____

____

Name ______________________ Date __________ Class __________

LESSON 8-10

Reteach

Transformations

A transformation moves a figure without changing its size or shape. So, a figure and its transformation are always congruent.

A translation is the movement of a figure along a straight line. The location is the only thing that changes with a translation.

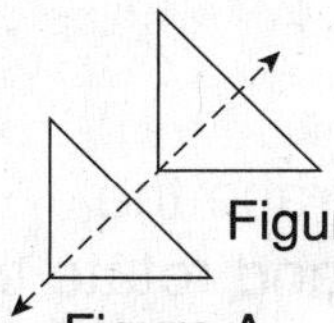

Figure B
Figure A

Figure B is a translation of Figure A.

A rotation is the movement of a figure around a point. The location and position of a figure can change with a rotation.

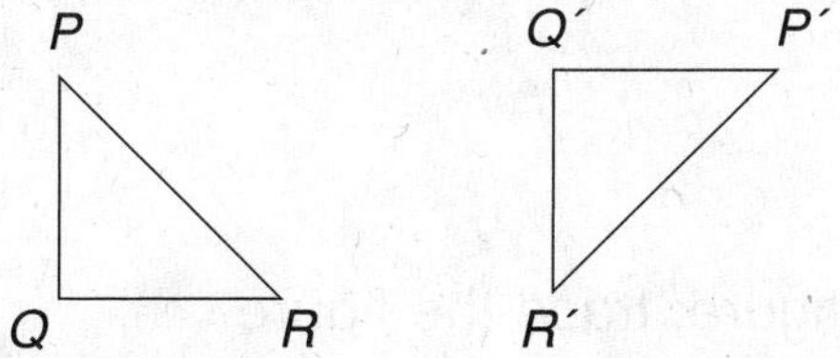

Figure C Figure D

Figure D is a rotation of Figure C around point *R*.

When a figure flips over a line creating a mirror image, it is called a reflection. The line the figure flips over is called the line of reflection. The location and position of a figure change with a reflection.

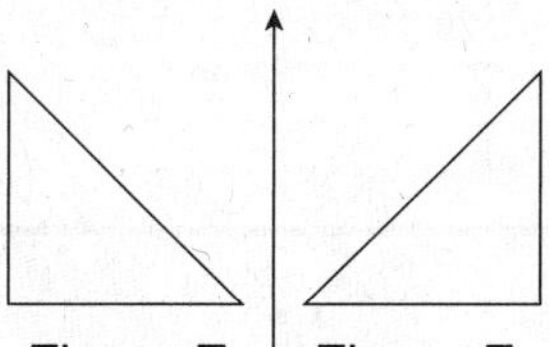

Figure E Figure F

Figure F is a reflection of Figure E.

Identify each transformation.

1.

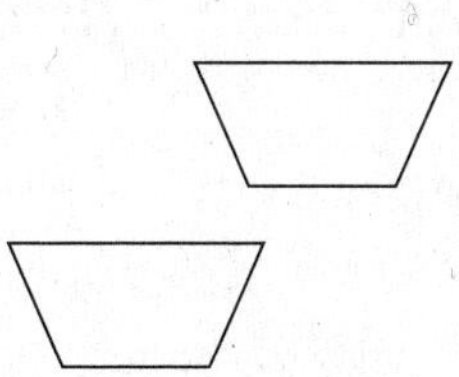

____

2.

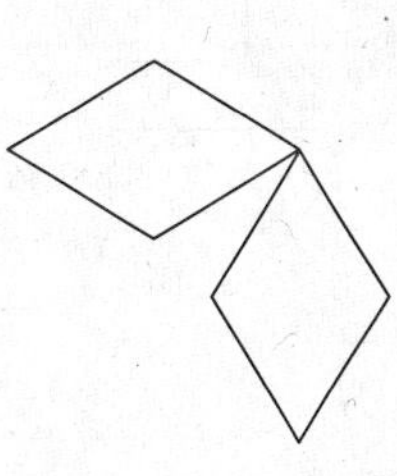

____

Name ______________________ Date ____________ Class ____________

LESSON 8-10
Reteach
Transformations (continued)

You can draw transformations.

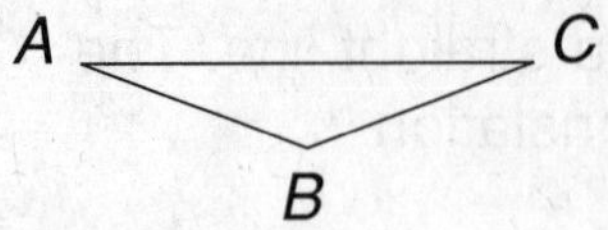

To draw a 45° clockwise rotation around point *C,* first trace the figure. Next, place your pencil on the point of rotation and rotate the figure clockwise 45°. Then trace the figure in its new location.

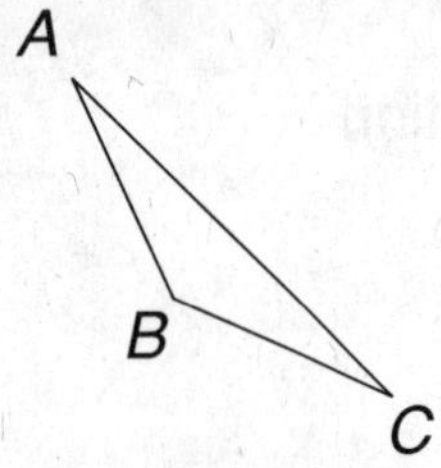

This is the original figure rotated 45° clockwise around point *C.*

To draw a vertical reflection of the original figure, trace the figure and flip it vertically over the line of reflection. Then trace the figure in its new location.

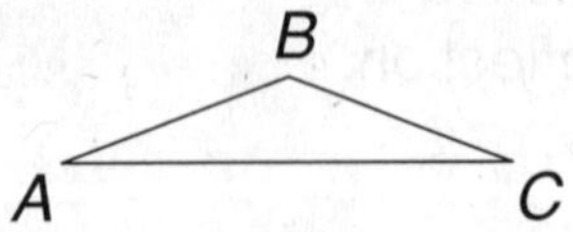

This is the vertical reflection of the original figure.

Draw each transformation.

3. vertical reflection

4. 60° clockwise rotation around point A

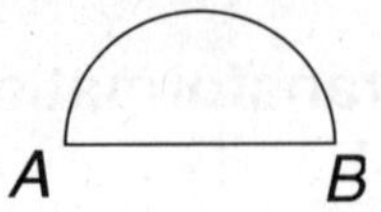

Name ______________________ Date ____________ Class ____________

LESSON **8-10**

Challenge
Rotation Ride

The Ferris wheel shown below is turning clockwise around its rotation point at the center of the wheel. For each person on the Ferris wheel, describe how many degrees he or she must rotate to be able to get off the ride.

1. Darius ____________________

2. Jenna ____________________

3. Paul ____________________

4. Sarah ____________________

5. Mei ____________________

6. Ben ____________________

7. Mindy ____________________

8. Carlos ____________________

Name ______________________ Date ____________ Class ____________

LESSON **8-10**
Problem Solving
Transformations

Write the correct answer.

1. If the rotation point of a circle is its center, how will all rotations affect the circle?

____

2. What transformation could make an arrow pointing east become an arrow pointing north?

____

3. What transformation could make the number 9 become the number 6?

____

____

4. What transformation could make the letter P look like the letter b?

____

5. On the coordinate plane at right, graph Triangle A with vertices (3, 1), (6, 1), and (3, 5). Then graph Triangle B with vertices (3, 6), (6, 6), and (3, 10). What transformation best describes the change from Triangle A to Triangle B?

____

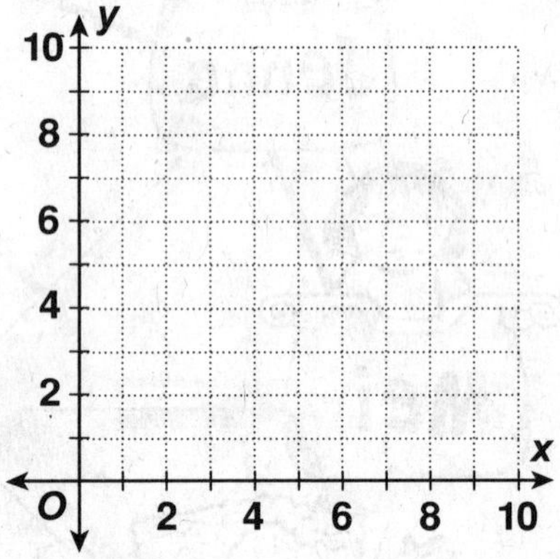

Circle the letter of the correct answer.

6. Which transformation best describes the figure on the right?

A 90° clockwise rotation
B horizontal reflection
C 90° counterclockwise rotation
D horizontal translation

7. Which transformation best describes the figure on the left?

F horizontal reflection
G 180° counterclockwise rotation
H 90° counterclockwise rotation
J horizontal translation

Name ______________________ Date ____________ Class ____________

LESSON **8-10**

Reading Strategies

Focus on Vocabulary

Moving a figure does not change its size or its shape.
Transformation means moving a figure. Here are three kinds of transformations.

- **Translate** → **Slide** a figure to a new position.

Letter *P* has slid from one spot to a new spot.

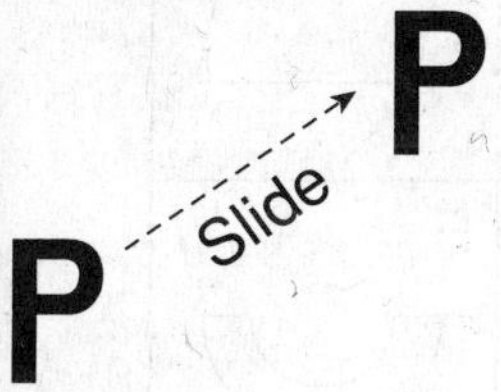

- **Rotate** → **Turn** a figure around a point to a new position.

Letter *P* has been rotated to a new spot.

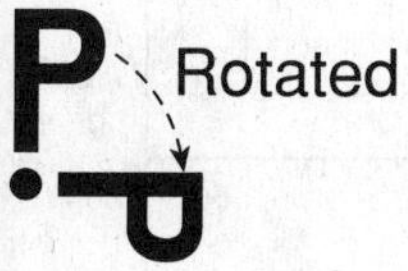

- **Reflect** → **Flip** a figure over a line to a new position.

Letter *P* has been flipped over a line of reflection.

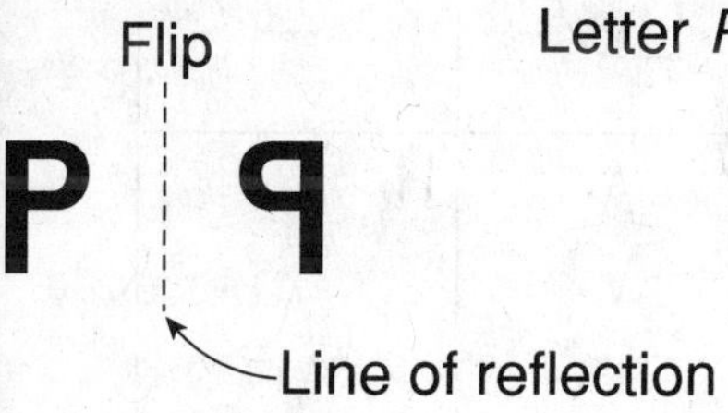

Write *translate, rotate,* or *reflect* to name the kind of transformation shown in each figure.

1.

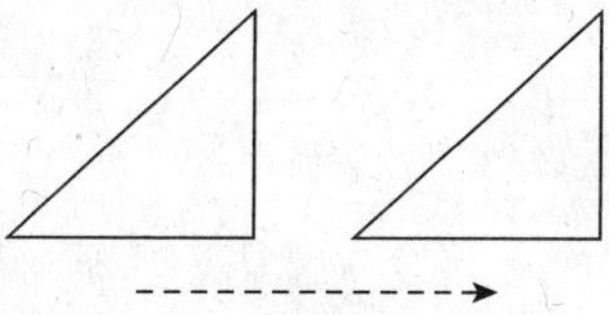

____

2.

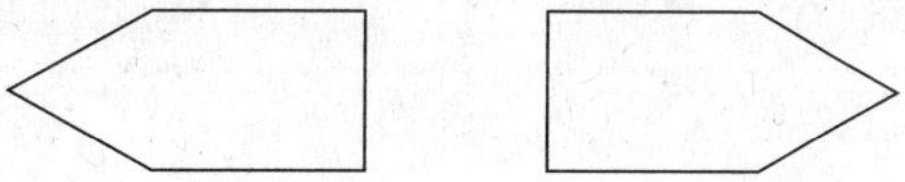

____

Name ______________________ Date ___________ Class ___________

LESSON 8-10

Puzzles, Twisters & Teasers

Transformed

How would you complete each statement? Shade the correct box.

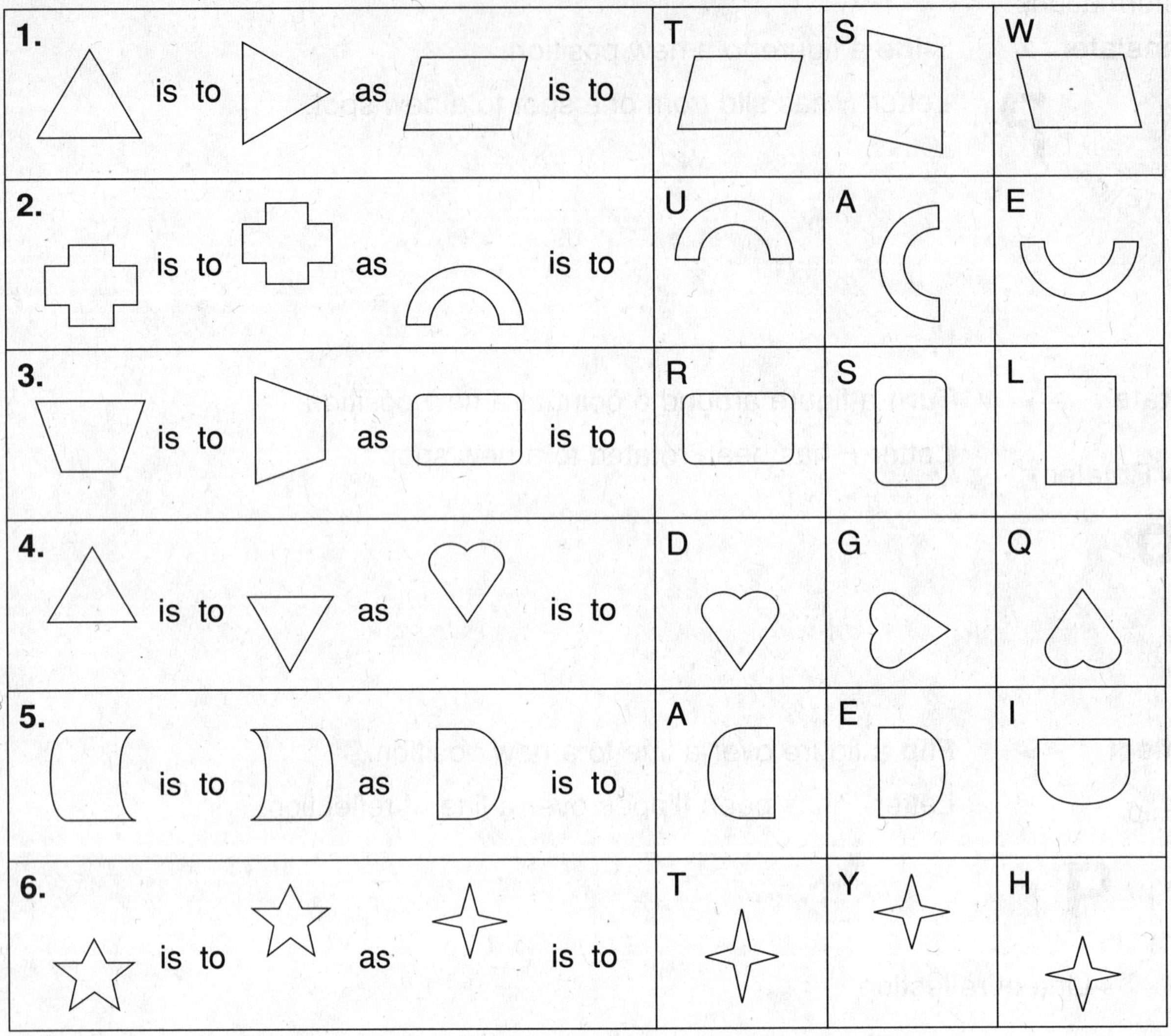

To solve the riddle, unscramble the letters in the shaded boxes.

What do you get when an elephant walks in your garden? ______ ________

Name ______________________ Date ____________ Class ____________

LESSON **8-11**

Practice A

Line Symmetry

Determine whether each dashed line appears to be a line of symmetry.

1.

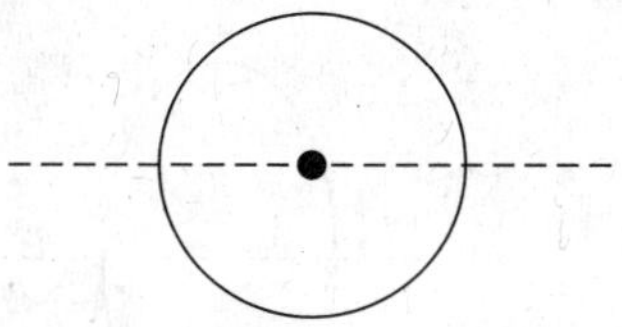

____

2.

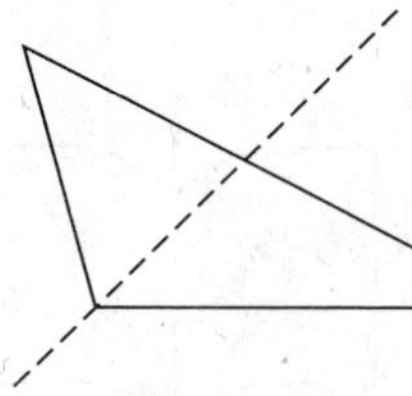

____

3.

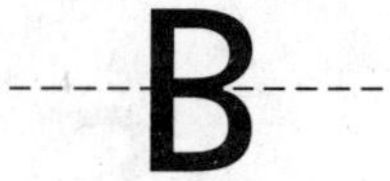

____

Find all of the lines of symmetry in each regular polygon.

4.

5.

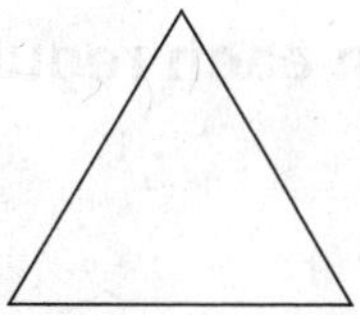

6.

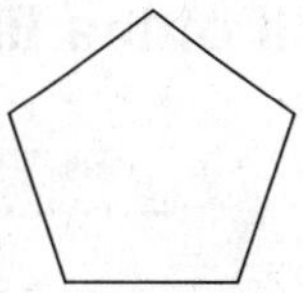

Draw each cut-out figure as it would look unfolded.

7.

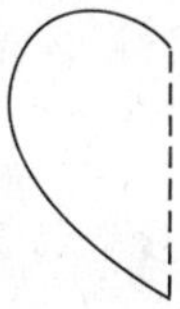

8.

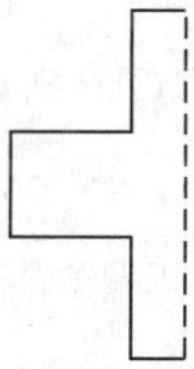

9. Draw a line of symmetry for the word.

DID

10. How do you know if a figure has line symmetry?

____

____

Name ______________________ Date ____________ Class ____________

LESSON 8-11

Practice B
Line Symmetry

Determine whether each dashed line appears to be a line of symmetry.

1.

____

2.

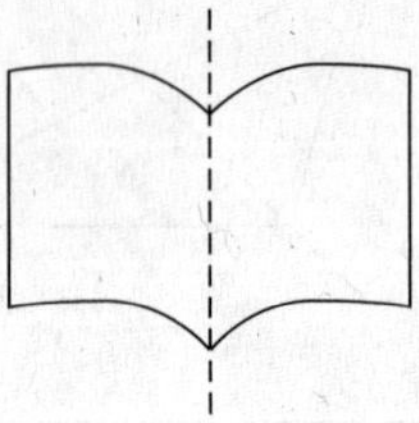

____

3.

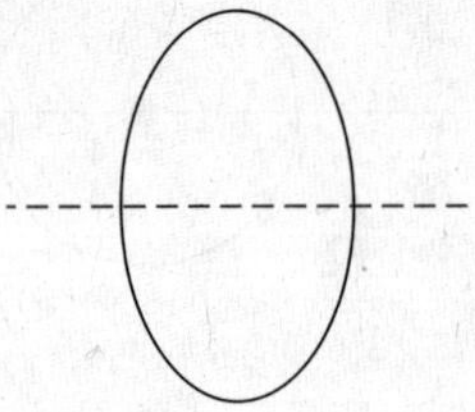

____

Find all of the lines of symmetry in each regular polygon.

4.

5.

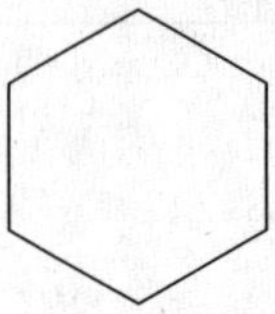

6.

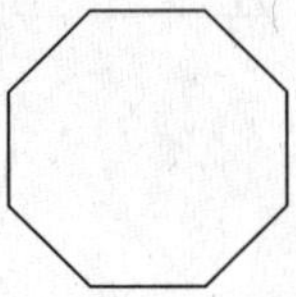

Draw each cut-out figure as it would look unfolded.

7.

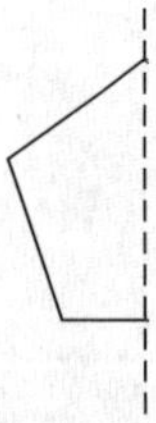

8.

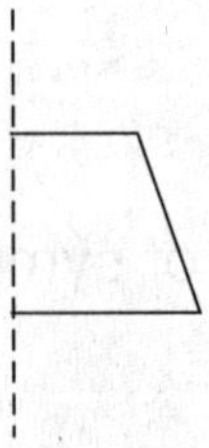

9. Which has more lines of symmetry, a square or a rectangle?

____

10. Of the numbers 1 through 9, which numbers can have lines of symmetry?

____

Name ______________________ Date ____________ Class ____________

LESSON 8-11

Practice C

Line Symmetry

Determine whether each dashed line appears to be a line of symmetry.

1.

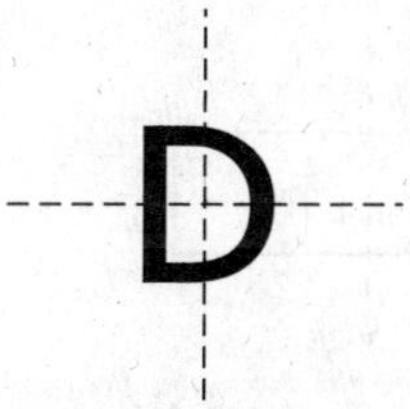

2.

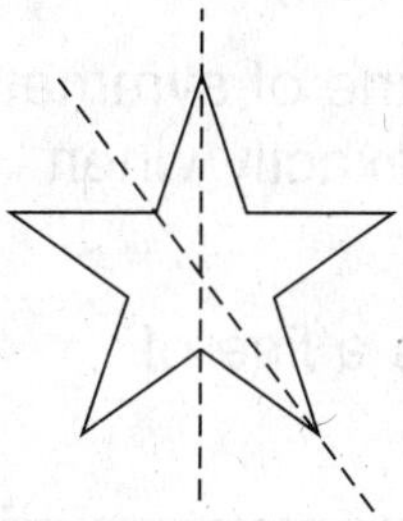

3.

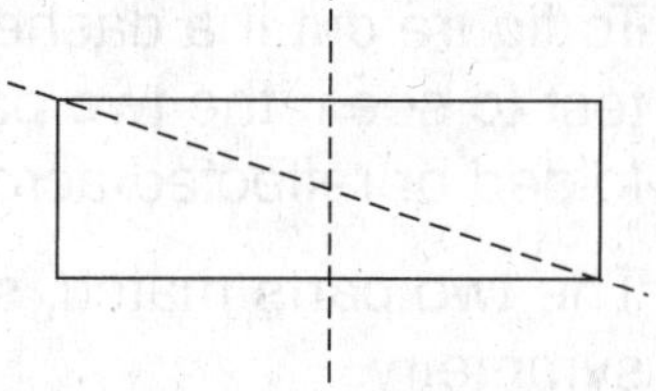

Find all of the lines of symmetry in each regular polygon.

4.

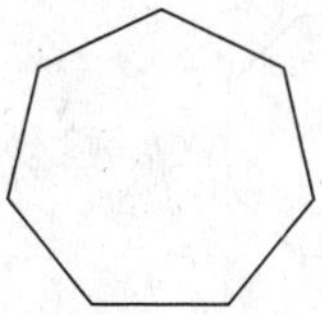

5.

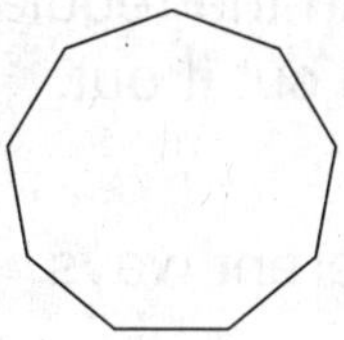

6.

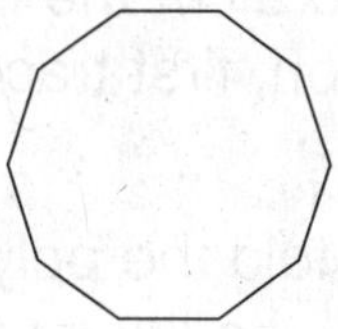

Draw each cut-out figure as it would look unfolded.

7.

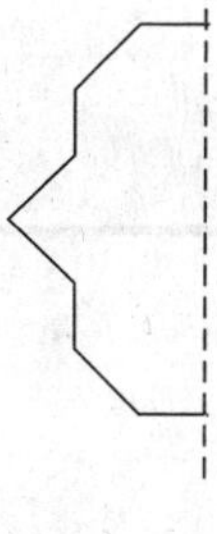

8.

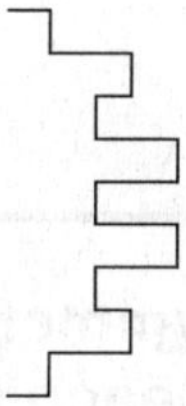

9. If all of the letters of a word have a horizontal line of symmetry through their centers, what would the word look like if it were vertically reflected?

Name ____________________ Date __________ Class __________

LESSON 8-11

Reteach

Line Symmetry

A figure has line symmetry if it can be folded or reflected so that the two overlapping parts of the figure are congruent. The line of reflection is called the line of symmetry.

To figure out if a dashed line is a line of symmetry, test to see if the two parts match exactly when folded or reflected across the line.

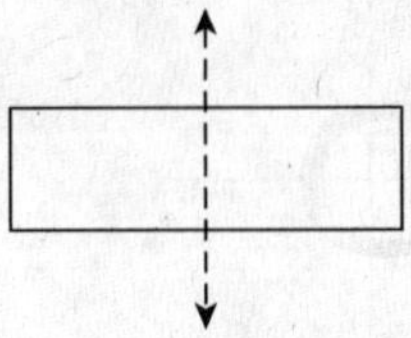

The two parts match, so the line is a line of symmetry.

Decide if each line is a line of symmetry.

1.

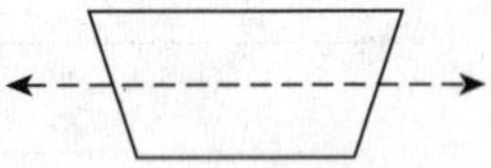

____

2.

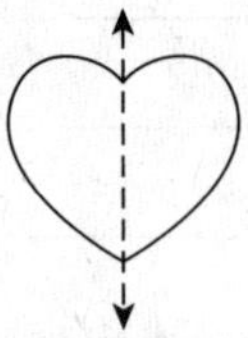

____

To find all of the lines of symmetry in the regular octagon, first trace the polygon and cut it out.

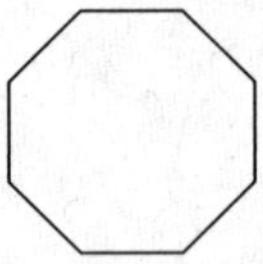

Then fold the polygon in half in different ways to find the lines of symmetry.

A regular octagon has 8 lines of symmetry.

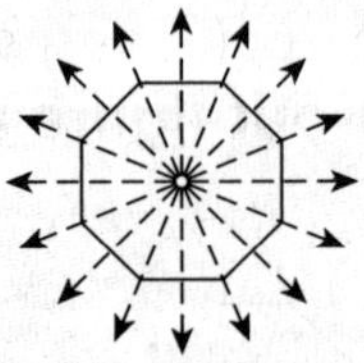

Find all of the lines of symmetry in each figure. Write how many lines of symmetry each figure has.

3.

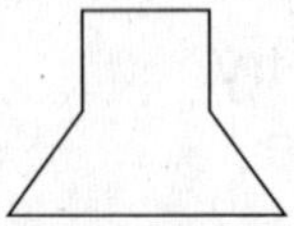

____

4.

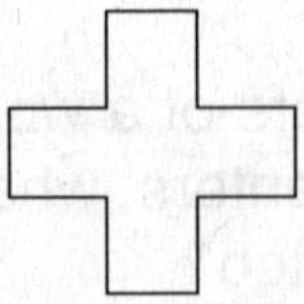

____

Name ______________________ Date ____________ Class ____________

LESSON **8-11**

Challenge

Alphabetical Symmetry

Tell whether each capital letter has a horizontal line of symmetry, a vertical line of symmetry, both, or neither. Draw the lines of symmetry to illustrate.

A	B	C	D
____________	____________	____________	____________
E	F	G	H
____________	____________	____________	____________
I	J	K	L
____________	____________	____________	____________
M	N	O	P
____________	____________	____________	____________
Q	R	S	T
____________	____________	____________	____________
U	V	W	X
____________	____________	____________	____________
Y	Z		
____________	____________		

Name ______________________ Date ____________ Class ____________

LESSON **8-11**

Problem Solving

Line Symmetry

Write the correct answer.

1. Do your body and face appear to have a vertical line of symmetry or a horizontal line of symmetry?

____

2. Which letter of the alphabet has an infinite, or endless, number of lines of symmetry?

____

3. Ted says the diagonals of a rectangle are also its lines of symmetry. Do you agree? Explain.

____

____

____

____

4. Using the digits 0 through 9 and not repeating any digits, write a 3-digit number that has a horizontal line of symmetry.

____

____

____

____

5. Draw a line of symmetry for this word.

DOCK

6. Draw the lines of symmetry for this star.

Circle the letter of the correct answer.

7. How many lines of symmetry does this hexagon have?

A 4
B 8
C 6
D 2

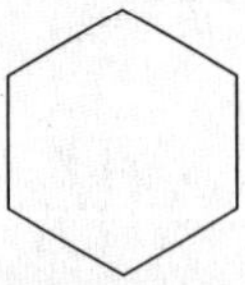

8. How many lines of symmetry does this flower have?

F 3
G 4
H 5
J 6

9. How many lines of symmetry does a square have?

A 0
B 2
C 4
D 6

10. How many lines of symmetry does a regular pentagon have?

F 1
G 2
H 4
J 5

Name ________________________________ Date ___________ Class ___________

LESSON 8-11

Reading Strategies

Focus On Examples and Non-Examples

If you fold a figure along a fold line and one side matches up exactly with the other side, the figure has **line symmetry.**

The fold line is called a **line of symmetry.**

Fold the rectangle along the diagonal line. The sides do not match up, so the fold line *is not* a line of symmetry.

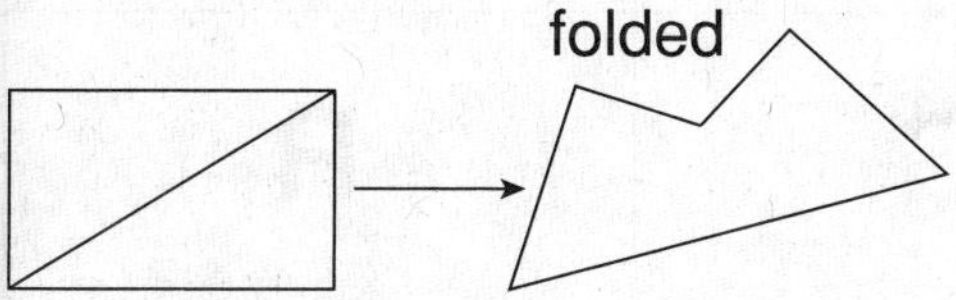

No Line Symmetry

Fold the rectangle along the vertical line. The sides match up exactly. So, the fold line *is* a line of symmetry.

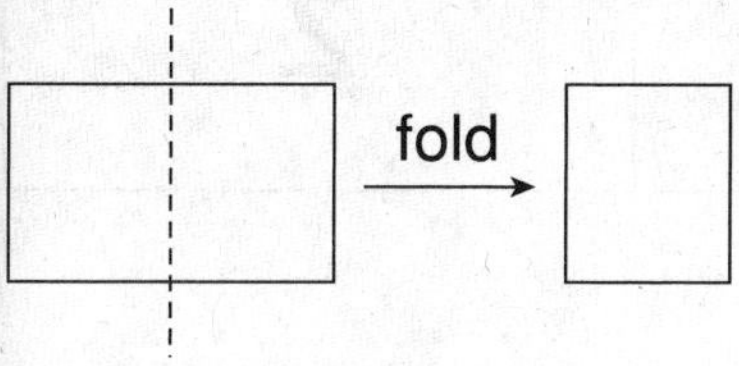

Line Symmetry

Trace and fold each figure below. If you can draw a line of symmetry, write *yes.* If not, write *no.*

1.

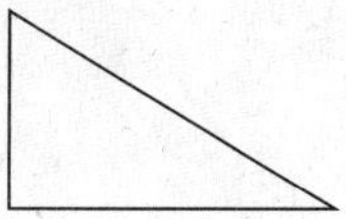

____

2.

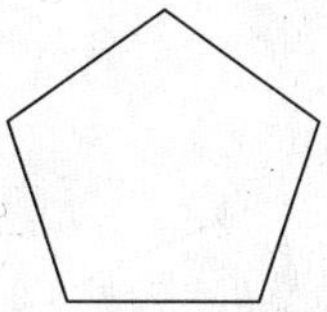

____

3.

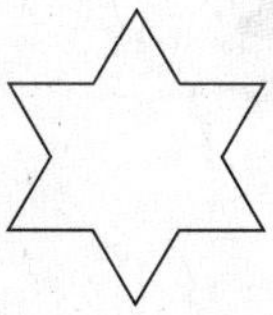

____

4.

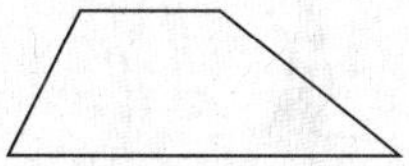

____

Name ______________________ Date __________ Class __________

LESSON 8-11

Puzzles, Twisters & Teasers

Mirror Images

Look at each letter. Circle the letters that have a line of symmetry, and then rearrange them to solve the riddle below.

N	E	D	P
I	A	G	R
T	L	S	Y
Q	F	M	Z

What is after sunrise and before sunset? ______________

LESSON 8-1 Practice A
Building Blocks of Geometry

Name each geometric figure.

1. •*P*

2. 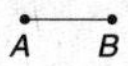

3. 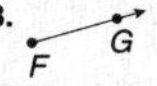

point *P*, or *P* $\overline{AB}$ or $\overline{BA}$ $\overrightarrow{FG}$

4. 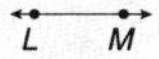

5. 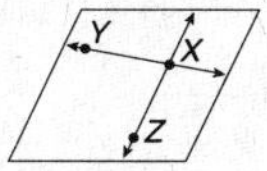

6. 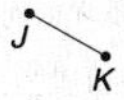

$\overleftrightarrow{LM}$ or $\overleftrightarrow{ML}$ plane *XYZ* $\overline{JK}$ or $\overline{KJ}$

Use the diagram to choose the correct answer.

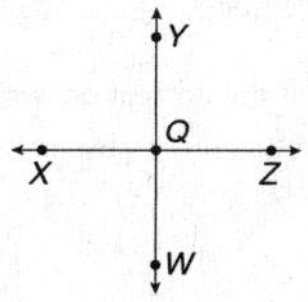

7. Which of the following does not name a line on the diagram?
- **A** Line *YQ*
- **(B)** Line *WZ*
- **C** $\overleftrightarrow{ZX}$
- **D** $\overleftrightarrow{QW}$

8. Which two points cannot be used to name a ray on the diagram?
- **F** *Q* and *Y*
- **G** *Z* and *X*
- **(H)** *X* and *Y*
- **J** *W* and *Q*

9. How many points are needed to define a line or line segment?

2 points

10. Lynn named this ray $\overrightarrow{PQ}$. Explain her error and give the correct name for the ray.

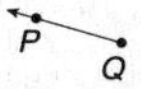

A ray must be named by its endpoint first.

The correct name is $\overrightarrow{QP}$.

LESSON 8-1 Practice B
Building Blocks of Geometry

Use the diagram to name each geometric figure. Possible answers are given.

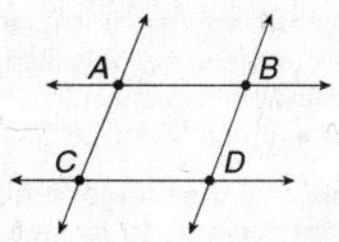

1. two points A and B

2. a plane plane *ABD*

3. a line segment $\overline{BD}$

4. a point shared by two lines A

5. a line $\overleftrightarrow{CD}$

Use the diagram to give a possible name to each figure. Possible answers are given.

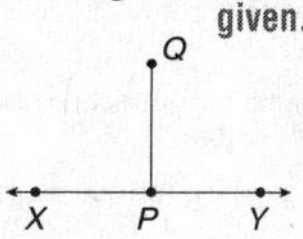

6. two different ways to name the line

line *XY* and $\overleftrightarrow{XY}$

7. four different names for rays

ray *PY*, ray *PX*, $\overrightarrow{PY}$, and $\overrightarrow{PX}$

8. another name for $\overline{QP}$

$\overline{PQ}$

9. Is the following statement always true, sometimes true, or never true? Explain your reasoning. A line is longer than a line segment.

It is always true, because a line segment only extends between two endpoints, but a line extends without end in opposite directions.

10. Using endpoints as your basis, explain how a line, a line segment, and a ray are different.

A line has no endpoints, a ray has one endpoint, and a line segment has two endpoints.

LESSON 8-1 Practice C
Building Blocks of Geometry

Use the diagram to name each geometric figure. Possible answers are given.

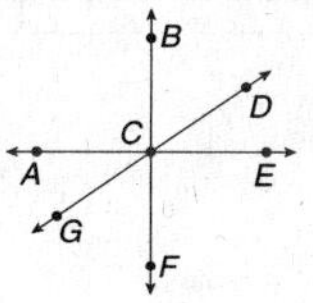

1. a point shared by three lines C

2. three points on the same line

A, C, and E

3. four different rays

$\overrightarrow{CD}$, $\overrightarrow{CE}$, $\overrightarrow{CF}$, and $\overrightarrow{CG}$

4. four different names for the same line segment

$\overline{AC}$, $\overline{CA}$, line segment *AC*, line segment *CA*

Draw and label an example of each geometric figure. Possible answers are given.

5. $\overrightarrow{PQ}$

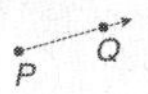

6. $\overleftrightarrow{ST}$

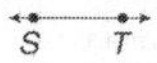

7. $\overline{XY}$

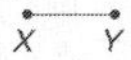

8. plane *FGH*

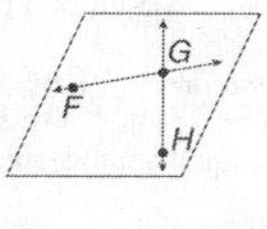

9. If 3 points are labeled on a line, how many different line segments can be named on that line? How many different rays?

3 different line segments and 6 different rays

10. Name every geometric figure that is part of line *AB*.

points *A* and *B*, $\overleftrightarrow{AB}$, $\overrightarrow{AB}$, $\overrightarrow{BA}$, and $\overline{AB}$

LESSON 8-1 Reteach
Building Blocks of Geometry

Here are terms that can help you understand geometry.

A point is an exact location. A point has no size.

•*A*   point *A* or *A*   Use a capital letter to name a point.

A line is a straight path that extends without end in opposite directions. A line has infinite length, but no width.

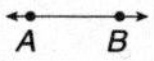

$\overleftrightarrow{AB}$ or $\overleftrightarrow{BA}$ Two points name a line.

A plane is a flat surface that extends without end in all directions. A plane has infinite length and width, but no depth.

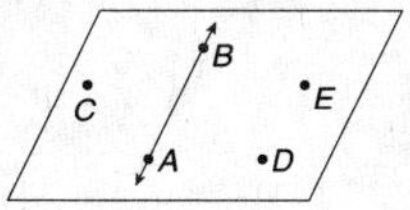

CDE *EDC*
DEC *CED*
ECD *DCE*

A plane is named by 3 points on the plane that are not on a line.

A line segment has two endpoints. The length of a line segment can be measured.

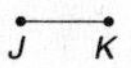

$\overline{JK}$ or $\overline{KJ}$ A line segment is named by its endpoints.

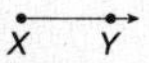

$\overrightarrow{XY}$

A ray has one endpoint and extends without end in one direction. A ray is named by its endpoint first and another point on the ray.

Use the diagram to name each geometric figure. Possible answers are given.

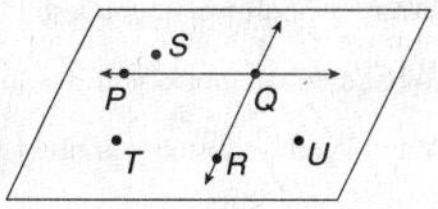

1. two points   R, S

2. two lines $\overleftrightarrow{PQ}$ and $\overleftrightarrow{RQ}$

3. a plane   STU

4. two line segments $\overline{PQ}$ and $\overline{QR}$

5. two rays $\overrightarrow{QP}$ and $\overrightarrow{QR}$

6. a point shared by two lines   Q

LESSON 8-1 Challenge
Points of Light

Astronomers have divided the sky into 88 constellations, or groups of stars. To map these constellations, astronomers use a point for each star and draw imaginary lines between the points. One of the easiest groups of stars to see is called the Big Dipper. The Big Dipper is not officially a constellation. It is only part of a large constellation called Ursa Major, or Great Bear.

Imagine the stars of Ursa Major are labeled as shown below. Follow these instructions to make a star map of Ursa Major.

1. Use line segments to connect points *A*–*G* in alphabetical order, and then draw $\overline{DG}$.
2. Use line segments to connect points *G*–*L* in alphabetical order, and then draw $\overline{FJ}$.
3. Use line segments to connect points *M*–*O* in alphabetical order. Then draw $\overline{EM}$ and $\overline{MP}$.
4. Label the star grouping formed by the 7 points *A*–*G* as the "Big Dipper."
5. Then label each point of the Big Dipper with its proper star name: *A*: Alkaid; *B*: Mizar; *C*: Alioth; *D*: Megrez; *E*: Phecda; *F*: Merak; *G*: Dubhe

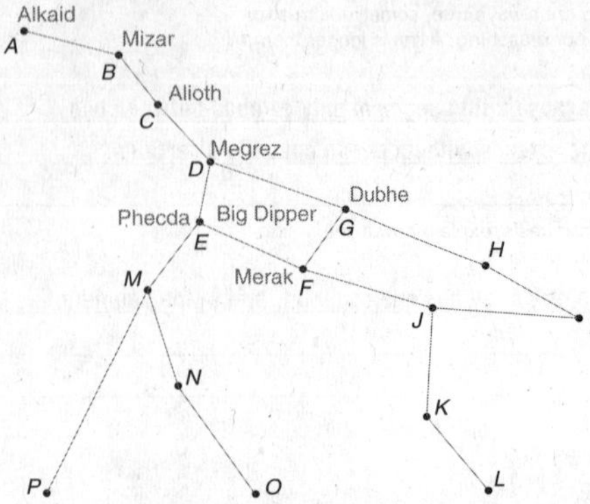

LESSON 8-1 Problem Solving
Building Blocks of Geometry

Place your hand down flat on a sheet of paper. Draw a point at the tip of your thumb, the tip of your middle finger, and the tip of your pinky.

1. Label the thumb point *A*, the middle finger point *B*, and the pinky point *C*.

   Check students' drawings.

2. Name all the planes you possibly can with points *A*, *B*, and *C*.

   plane *ABC*

3. Draw and name all the lines you can make with points *A*, *B*, and *C*.

   $\overleftrightarrow{AB}$, $\overleftrightarrow{AC}$, and $\overleftrightarrow{BC}$

4. Name all the line segments possible using points *A*, *B*, and *C*.

   $\overline{AB}$, $\overline{AC}$, and $\overline{BC}$

5. Name all the rays possible using points *A*, *B*, and *C*.

   $\overrightarrow{AB}$, $\overrightarrow{AC}$, $\overrightarrow{BA}$, $\overrightarrow{BC}$, $\overrightarrow{CA}$, and $\overrightarrow{CB}$

6. Choose one line that you drew. Give all the different possible names for that line.

   Possible answer: $\overleftrightarrow{AB}$, $\overleftrightarrow{BA}$, line *AB*, and line *BA*

Circle the letter of the correct answer.

7. Which of the following has exactly one endpoint?
   - A $\overline{OP}$
   - B $\overleftrightarrow{AB}$
   - C $\overline{TR}$
   - (D) $\overrightarrow{SM}$

8. Which of the following is a straight path that extends without end in opposite directions?
   - F a point
   - (G) a line
   - H a ray
   - J a line segment

9. Which statement is false?
   - A An infinite number of lines can be drawn through one point.
   - B Exactly one line can be drawn between two points.
   - (C) A line contains exactly one ray.
   - D If points *A* and *B* are on a line, then line segment *AB* and line segment *BA* are the same.

10. Why is the false statement in Exercise 9 not true?
    - (F) Any point on a line defines another ray on the line.
    - G A line contains exactly two rays.
    - H A line contains exactly five rays.
    - J A line does not contain any rays.

LESSON 8-1 Reading Strategies
Understanding Symbols

A **point** marks an exact location in space. Read: "point *T*."

A capital letter is used to name a point.

A **line** is a straight path that extends forever in both directions. Read: "line *DR*" or "line *RD*."

Two points on the line name a line.

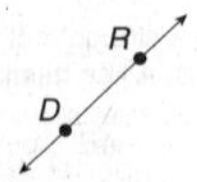

A **ray** is part of a line that has one endpoint and goes on forever in one direction. Read: "ray *SM*."

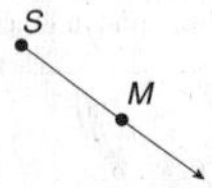

A **line segment** is part of a line between two endpoints.

Use the two endpoints to name the line.
Read: "segment *VW*" or "segment *WV*."

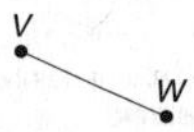

Use this figure to answer the questions.

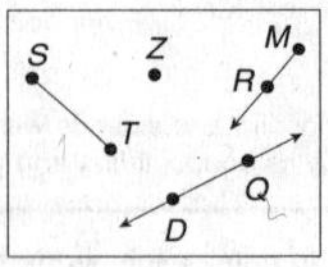

1. What symbols would you use to name the line segment? $\overline{ST}$ or $\overline{TS}$
2. What symbols would you use to name the line? $\overleftrightarrow{DQ}$ or $\overleftrightarrow{QD}$
3. What symbols would you use to name the ray? $\overrightarrow{MR}$
4. Write the name of the point. point *Z*
5. What is the difference between a line and a ray?

   Possible answer: A line is a path that goes forever in both directions. A ray has one endpoint and goes forever in one direction.

LESSON 8-1 Puzzles, Twisters & Teasers
Is it the Truth?

Determine which statements are true. Rearrange the letters of the true statements to solve the riddle.

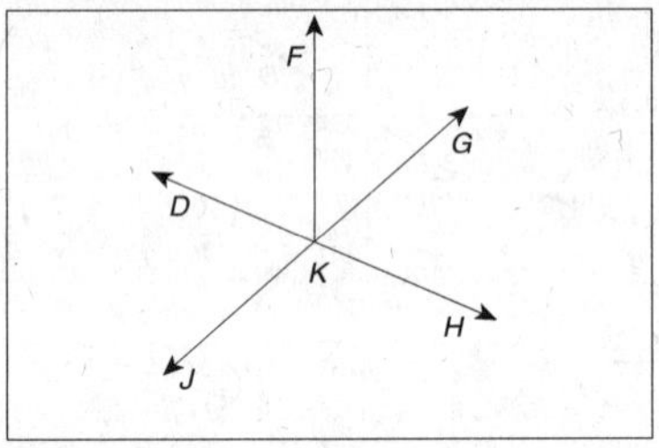

1. [R] Point *G* is on $\overleftrightarrow{JG}$. True
2. [O] $\overleftrightarrow{DH}$ can also be written $\overleftrightarrow{HD}$. True
3. [B] *JKG* is a plane.
4. [S] *GK* is a ray.
5. [A] A point is named by a capital letter. True
6. [N] *KD* is a ray. True
7. [U] A line segment has only one end point.
8. [T] *FKG* is a plane. True
9. [C] A line is a straight path that extends without end in all directions. True
10. [O] A ray is named by its endpoint first. True

What do you call a song played on an automobile radio?

A C A R T O O N

LESSON 8-2

Practice A

Measuring and Classifying Angles

Name the type of angle shown.

1. 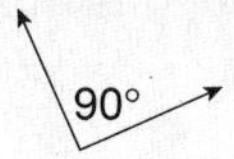

right angle

2. 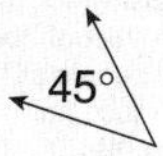

acute angle

3. 

straight angle

4. 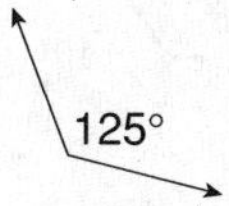

obtuse angle

5. 

right angle

6. 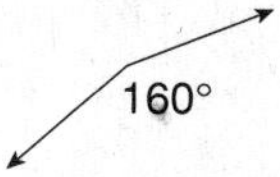

obtuse angle

Use a protractor to measure each angle.

7. 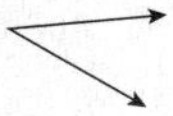
35°

8. 
90°

9. 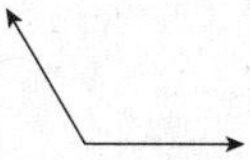
120°

10. Carl is making a desk for his office. One of the angles of the desk measures greater than 90°. What two types of angles could that angle be?

an obtuse angle or a straight angle

11. How many angles make up the capital letter L? What types of angles are they?

one right angle, or two straight angles

LESSON 8-2

Practice B

Measuring and Classifying Angles

Use a protractor to measure each angle.

1. 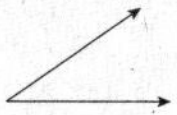
35°

2. 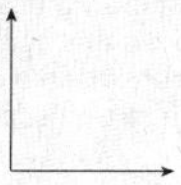
90°

3. 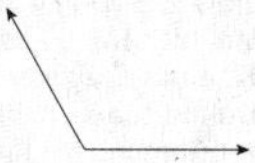
120°

Use a protractor to draw an angle with each given measure.

4. 70° 

5. 115° 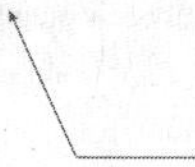

6. 45° 

Classify each angle as acute, right, obtuse, or straight.

7. 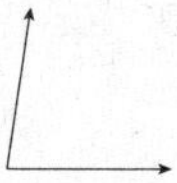
acute

8. 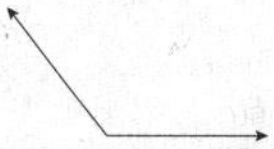
obtuse

9. 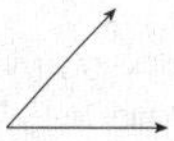
acute

10. The frame for a kite has two angles that together form a right angle. What type of angle is each of those angles? Explain.

They are both acute angles; The sum of the two angles is a right angle, so they each must measure less than 90°.

11. What kinds of angles are in each of the letters in this word? **TAXI**

The *T* has all right angles; the *A* has acute and obtuse angles; the *X* has acute and obtuse angles; and the *I* has one straight angle.

LESSON 8-2

Practice C

Measuring and Classifying Angles

Use a protractor to measure each angle and name the type of angle.

1. 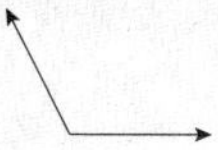
117°

2. 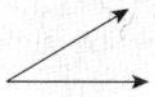
31°

3. 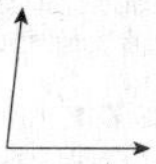
84°

Use a protractor to draw each angle. Possible answers are given.

4. an acute angle whose measure is greater than 45°
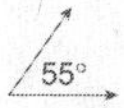

5. an obtuse angle whose measure is between 95° and 120°
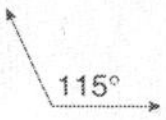

6. an obtuse angle formed by a right angle and an acute angle
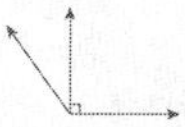

Classify each angle as acute, right, obtuse, or straight.

7. 
obtuse

8. 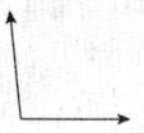
obtuse

9. 
acute

10. Together, ∠1 and ∠2 form a straight angle. If ∠1 is acute, what must be true about ∠2? Explain.

It must be obtuse; A straight angle measures 180°, so $\angle 1 + \angle 2 = 180°$. Because ∠1 is less than 90°, ∠2 must be greater than 90°.

11. Is the following statement always true, sometimes true, or never true? Explain your reasoning. Two acute angles of the same size form a right angle.

It is sometimes true. When the two acute angles measure 45°, they do form a right angle. However, if they do not measure 45°, they could form another acute angle or an obtuse angle.

LESSON 8-2

Reteach

Measuring and Classifying Angles

An angle is formed by two rays with a common endpoint, called the vertex.

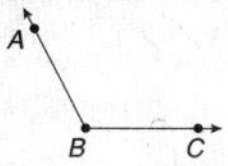

You can name an angle in different ways.

By its vertex ∠*B* OR By its vertex and one point from each ray. ∠*ABC*, ∠*CBA* Vertex letter in the middle.

Give three ways to name each angle.

1. 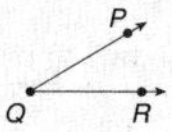

∠*Q*, ∠*PQR*, ∠*RQP*

2. 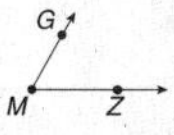

∠*M*, ∠*GMZ*, ∠*ZMG*

You can use a protractor to measure an angle.

To measure the angle, place the center point of the protractor on the vertex, *K*. Make sure that the protractor is placed so that $\overline{KL}$ passes through 0°. The measure of the angle is where $\overline{KJ}$ crosses the protractor.

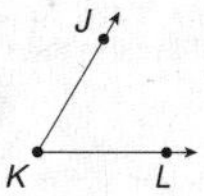

The measure of the angle is 60 degrees.

Write m∠*K* = 60°.

Use a protractor to measure each angle.

3. 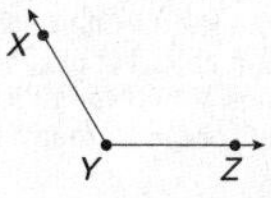

m∠*Y* = 120°

4. 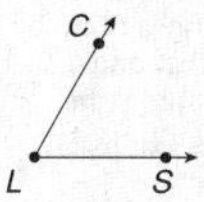

m∠*L* = 60°

LESSON 8-2
Reteach
Measuring and Classifying Angles (continued)

You can use a protractor to draw an angle.

To draw ∠*PQR* that measures 145°, first draw a base ray. Label the endpoint *Q*. Next, arrange the protractor so that its center is on *Q* and the base ray passes through 0°. Then, make a mark at 145° above the scale on the protractor. Use a straightedge to draw a ray from *Q* through the mark you made.

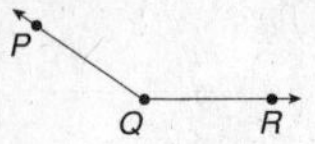

Label one point on a ray *P* and a point on the other ray *R*.

Use a protractor to draw an angle with each given measure.

5. 50°

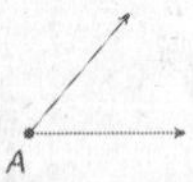

6. 170°

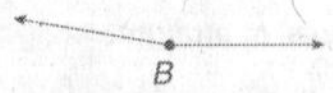

7. 85°

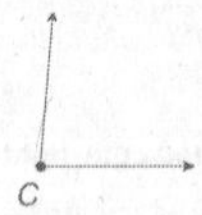

8. 110°

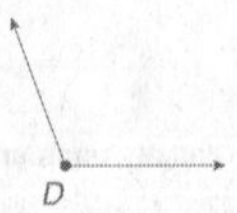

You can classify an angle by its measure.

If the measure is:	< 90°	> 90°	= 90°	= 180°
Then the angle is:	acute	obtuse	right	straight

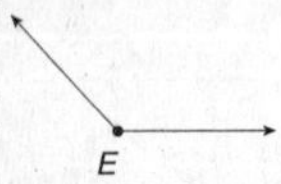

This angle is an obtuse angle because its measure is greater than 90°

Classify each angle.

9.

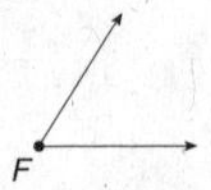

acute

10.

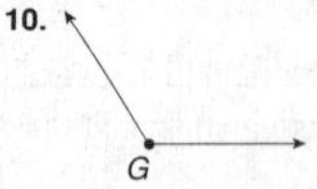

obtuse

LESSON 8-2
Challenge
I'll Meet You in 30°

No one knows for sure when people invented clocks to measure time. The first clocks were probably sundials, which measure time by the sun's shadow moving around a circle. Although today's clocks no longer use the sun to measure time, many people still use a circle to display time. A circle has 360°, and a clock face has 12 hours marked. So, each hour on the clock represents 30° of the circle.

Write the time shown on each clock below. Then use a protractor to measure each angle formed by the clock's hands and tell what kind of angle it is.

1.

3:00; 90°; right angle

2.

11:25; 180°; straight angle

3.

9:05; 120°; obtuse angle

4.

5:10; 90°; right angle

5.

8:00; 120°; obtuse angle

6.

4:30; 60°; acute angle

7.

2:35; 150°; obtuse angle

8.

6:00; 180°; straight angle

9.

10:50; 30°; acute angle

LESSON 8-2
Problem Solving
Measuring and Classifying Angles

Write the correct answer.

1. When a patient is lying flat in a hospital bed, what type of angle does the patient's body form? What is the measurement of that angle?

straight angle; 180°

2. When a patient is sitting straight up in a hospital bed, the upper body has been raised to what angle? What type of angle is that?

90° angle; right angle

3. Most hospital beds have a setting for the Fowler position. In this position, the patient's upper body is raised to form a 60° to 70° angle from a flat position. What types of angles are these?

They are both acute.

4. What are the greatest and least differences between the straight-up position and the Fowler position in a hospital bed?

least: 20°; greatest: 30°

Circle the letter of the correct answer.

5. Medical technicians often set the handles of crutches so that the patient's elbow is at a 30° angle. What type of angle is this?

(A) acute angle
B right angle
C obtuse angle
D straight angle

6. By law, wheelchair ramps in public places cannot be greater than 5 degrees. Which type of angle does a wheelchair ramp in public form with the ground?

(F) acute angle
G right angle
H obtuse angle
J straight angle

7. Physical therapists use a goniometer to measure the extension of a sitting patient's knee. Resting is 90°, and full extension is 180°. What angle does the goniometer measure if the patient's knee is at $\frac{1}{2}$ extension?

A 45°
B 90°
(C) 135°
D 0°

8. The Q-angle is measured between two points on a patient's hip joint and one point on the knee joint. A normal Q-measure for men is 14° plus or minus 3 degrees. What type of angle is any normal Q-angle for men?

F straight
G obtuse
H right
(J) acute

LESSON 8-2
Reading Strategies
Use Graphic Aids

Two rays with the same endpoint form an **angle.** The endpoint is called the **vertex.** Three letters can name an angle—a point on each side and the vertex as the middle letter. A single letter, the vertex, can also name it.

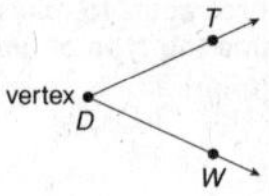

Read: "angle *TDW*," "angle *WDT*," or "angle *D*."

The size or measure of an angle can also name it.

A **right angle** measures exactly 90°. The symbol for a right angle is ⌐.

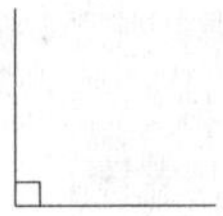

An **acute angle** measures less than 90°.

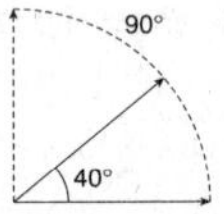

An **obtuse angle** measures more than 90°.

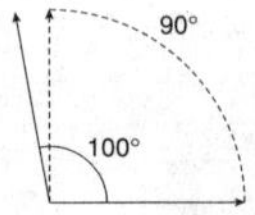

Write *right, acute,* or *obtuse* to name each angle.

1.

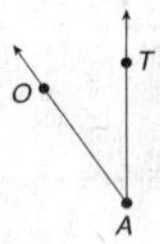

acute

2.

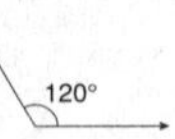

obtuse

3.

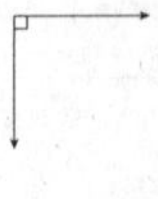

right

4.

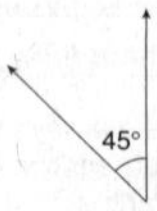

acute

LESSON 8-2
Puzzles, Twisters & Teasers
What's my Angle?

Measure each angle. Write the letter naming the angle on the line above the correct answer at the bottom of the page. Some letters will not be used.

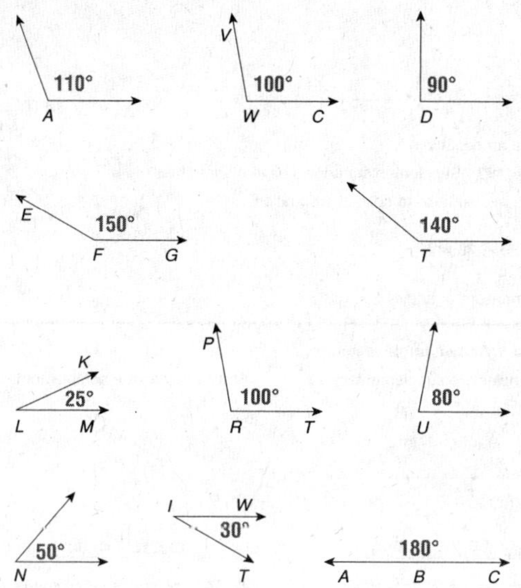

What do you call a boring fish?

A	D	U	L	L	-	F	I	N
110°	90°	80°	25°	25°		150°	30°	50°

LESSON 8-3
Practice A
Angle Relationships

Choose the correct answer.

1. Which best describes the type of angle pair shown below?

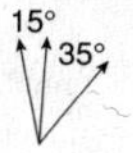

A vertical angles
(B) adjacent angles
C supplementary angles
D complementary angles

2. Which best describes the type of angle pair shown below?

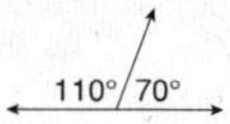

F vertical angles
G adjacent angles
(H) supplementary angles
J complementary angles

3. The angle pair shown below are complementary angles. What is the unknown angle measurement?

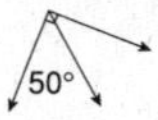

A 50°
(B) 40°
C 130°
D 90°

4. The angles shown below are vertical angles. What is the unknown angle measurement?

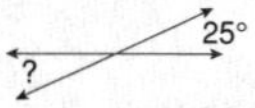

(F) 25°
G 65°
H 155°
J 50°

5. Two adjacent angles form a straight angle. How else can you describe the two angles?

They are supplementary angles.

6. Kelly subtracts 60° from 90° to find the measure of a missing angle in an angle pair. What type of angle pair is she working with?

complementary angles

LESSON 8-3
Practice B
Angle Relationships

Identify the type of each angle pair shown.

1.

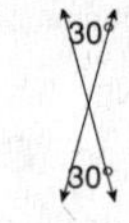

vertical angles

2.

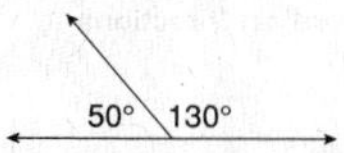

supplementary angles

3.

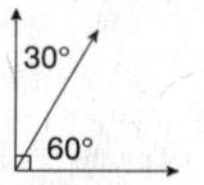

complementary angles

4.

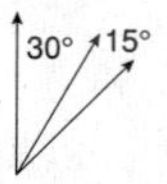

adjacent angles

Find each unknown angle measure.

5. The angles are supplementary.

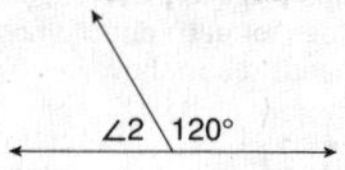

$m\angle 2 = 60°$

6. The angles are complementary.

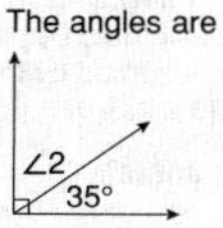

$m\angle 2 = 55°$

7. Anita says the plus sign + forms 2 pairs of vertical angles. Charles says it forms 2 pairs of congruent angles. Who is correct? Explain.

They are both correct. The intersecting lines of the plus sign form pairs of vertical angles. Vertical angles always have the same measure, so they are always congruent.

8. Is the following statement always true, sometimes true, or never true? Explain your reasoning. Two congruent angles that are complementary both measure 45°.

It is always true. Congruent angles have the same measure, and the sum of two complementary angles is 90°; 90° ÷ 2 = 45°.

LESSON 8-3
Practice C
Angle Relationships

Use the figure for Exercises 1–4. **Possible answers are given.**

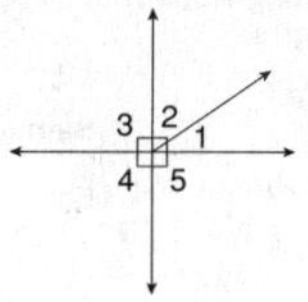

1. Name a pair of adjacent angles.

∠2 and ∠3

2. Name a pair of vertical angles.

∠3 and ∠5

3. Name a pair of complementary angles.

∠1 and ∠2

4. Name a pair of supplementary angles.

∠4 and ∠5

Find each unknown angle measure.

5. ∠1 and ∠2 are vertical angles. ∠1 measures 48°.

$m\angle 2 = 48°$

6. ∠3 and ∠4 are supplementary angles. ∠4 measures 117°.

$m\angle 3 = 63°$

7. ∠7 and ∠8 are complementary angles. ∠7 measures 23°.

$m\angle 8 = 67°$

8. ∠5 and ∠6 are congruent angles. ∠6 measures 12.5°.

$m\angle 5 = 12.5°$

9. If ∠1 and ∠2 are supplementary angles, and ∠3 and ∠2 are supplementary angles, what do you know about ∠1 and ∠3? Explain.

The measure of ∠1 equals the measure of ∠3. $m\angle 1 + m\angle 2 = 180°$, so $m\angle 1 = 180° - m\angle 2$. $m\angle 3 + m\angle 2 = 180°$, so $m\angle 3 = 180° - m\angle 2$. Therefore, $m\angle 3 = m\angle 1$.

10. Two angles are complementary. One angle is 20° smaller than the other angle. What are the measures of both angles?

35° and 55°

LESSON 8-3

Reteach

Angle Relationships

When two lines intersect, two pairs of opposite angles, called vertical angles, are formed. Vertical angles are congruent.

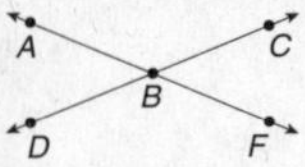

∠*ABC* and ∠*DBF* are vertical angles.
∠*ABD* and ∠*CBF* are vertical angles.

Adjacent angles have a common side and vertex.

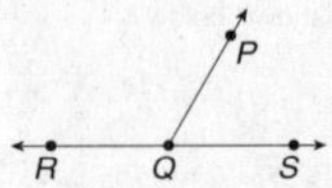

∠*PQR* and ∠*PQS* are adjacent angles.

Identify the type of each angle pair shown.

1.

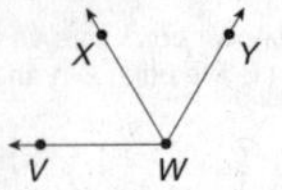

adjacent angles

2.

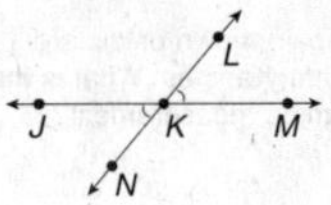

vertical angles

If the sum of the measures of two angles is 90°, then the angles are complementary angles.

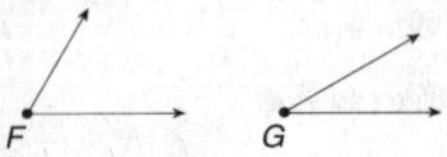

∠*F* and ∠*G* are complementary angles.

If the sum of the measures of two angles is 180°, then the angles are supplementary angles.

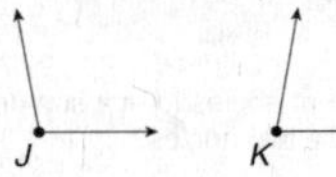

∠*J* and ∠*K* are supplementary angles.

Identify the type of each angle pair shown.

3.

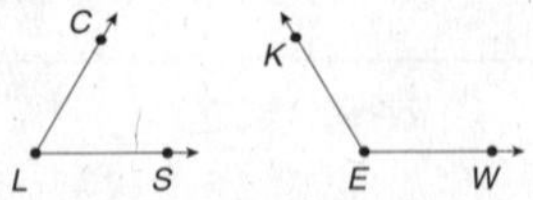

supplementary angles

4.

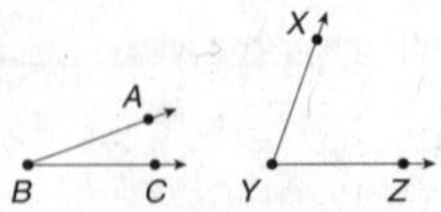

complementary angles

LESSON 8-3

Reteach

Angle Relationships (continued)

You can use related facts to find an unknown angle measure.

If you know that angles *J* and *K* are supplementary and m∠*J* = 78°, you can find the measure of angle *K*.

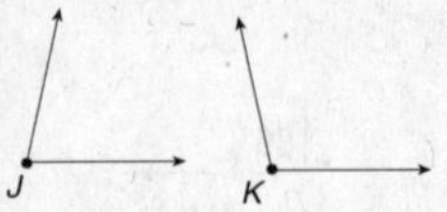

First write an equation.

78 + *K* = 180 Supplementary angles have a sum of 180°.

Then use related facts to solve the equation.

78 + *K* = 180
Think: 180 − 78 = *K*
K = 102

So, the measure of angle *K* = 102°.

Find each unknown angle measure.

5. The angles are supplementary.

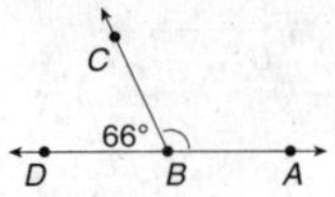

m∠*ABC* = 114°

6. The angles are complementary.

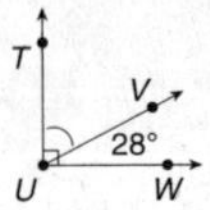

m∠*TUV* = 62°

7. The angles are vertical angles.

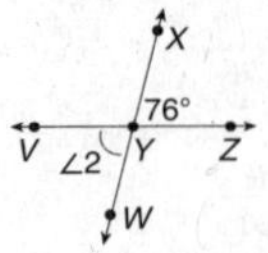

m∠2 = 76°

8. The angles are supplementary.

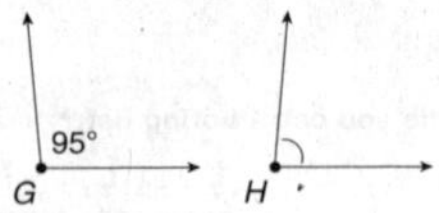

m∠*H* = 85°

LESSON 8-3

Challenge

Angle Partners

Measure all the angles in the chart below. Then draw lines to match each starting angle with its complementary and supplementary angles.

Complementary Angles	Starting Angles	Supplementary Angles
11°	63°	165°
9°	15°	117°
48°	81°	99°
27°	42°	101°
75°	79°	138°

LESSON 8-3

Problem Solving

Angle Relationships

Use the two compass roses to answer questions 1-6.

Cardinal Directions

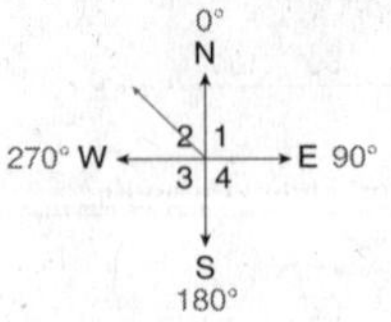

Intermediate Directions

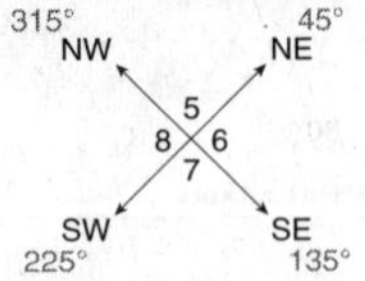

1. Which angles formed by the cardinal directions are vertical angles?

∠1 and ∠3; ∠2 and ∠4

2. Which angles formed by the intermediate directions are vertical angles?

∠5 and ∠7; ∠6 and ∠8

3. Draw the northwest directional ray on the cardinal compass rose. Describe the adjacent angles formed by the new ray. **Possible answer:**

complementary angles

4. North on a compass is 0°, and east is 90°. Use this information to label the degrees for each direction on the two compass roses above.

Circle the letter of the correct answer.

5. Which angles formed by the cardinal directions are supplementary to ∠2?
A ∠1
B ∠1 and ∠3
C ∠3 and ∠4
(D) ∠1, ∠3 and ∠4

6. Which angles formed by the intermediate directions are supplementary to ∠6?
F ∠5
G ∠5 and ∠7
H ∠7 and ∠8
(J) ∠5, ∠7 and ∠8

7. Angles *A* and *B* are complementary. ∠*B* is twice as large as ∠*A*. What are the measurements for each angle?
A ∠*A* = 45°; ∠*B* = 90°
(B) ∠*A* = 30°; ∠*B* = 60°
C ∠*A* = 60°; ∠*B* = 120°
D ∠*A* = 90°; ∠*B* = 180°

8. ∠1 and ∠2 are complementary. ∠2 and ∠3 are supplementary. The measure of ∠1 is 45°. What is the measure of ∠3?
F 45°
G 270°
H 90°
(J) 135°

LESSON 8-3
Reading Strategies
Draw Conclusions

Congruent means exactly the same size and shape.
Angles that have the same measure are congruent.

Vertical angles are angles that are opposite each other.
When two lines intersect, vertical angles are formed.

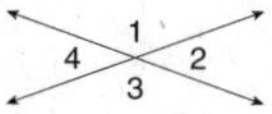

∠1 and ∠3 are vertical angles.

∠2 and ∠4 are vertical angles.

- Vertical angles have the same measure.
- Vertical angles are congruent.

Adjacent angles sit next to each other, or side-by-side. They share a common vertex and ray. Adjacent angles may or may not be congruent.

In the diagram above, ∠1 and ∠2 are adjacent angles. ∠4 and ∠3 are also adjacent angles.

Use this figure to answer the following questions. Use *congruent, vertical,* and *adjacent* to describe each angle. There may be more than one correct answer.

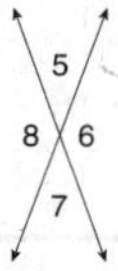

1. Study ∠5 and ∠6. What can you conclude about these two angles?

   They are adjacent.

2. Study ∠5 and ∠7. What can you conclude about these two angles?

   They are vertical and congruent.

3. Study ∠6 and ∠8. What can you conclude about these two angles?

   They are vertical and congruent.

LESSON 8-3
Puzzles, Twisters & Teasers
Pick a Pair

Circle the pairs of angles that answer each question. Rearrange the letters from the unused angles to answer the riddle.

1. Which 2 pairs of angles are supplementary?

56°	92°	37°	118°	124°	62°	53°
K	N	E	Y	G	Q	L

2. Which 2 pairs of angles are vertical?

83°	42°	48°	126°	97°	126°	42°
J	K	A	N	L	O	R

3. Which 2 pairs of angles are complementary?

33°	109°	71°	76°	57°	104°	19°
W	Y	P	E	V	B	H

What kind of bean does not grow from seed? jelly bean

LESSON 8-4
Practice A
Classifying Lines

Choose the correct answer.

1. Which relationship best describes the pair of lines shown below?

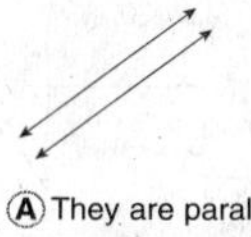

   (A) They are parallel lines.
   B They are intersecting lines.
   C They are perpendicular lines.
   D They are skew lines.

2. Which relationship best describes the pair of lines shown below?

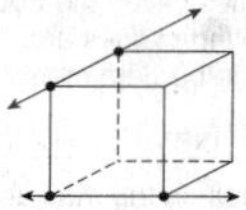

   F They are parallel lines.
   G They are intersecting lines.
   H They are perpendicular lines.
   (J) They are skew lines.

3. Which relationship best describes the pair of lines shown below?

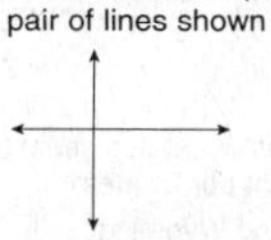

   A They are parallel lines.
   B They are intersecting lines.
   (C) They are perpendicular lines.
   D They are skew lines.

4. Which relationship best describes the pair of lines shown below?

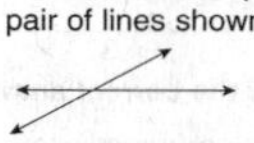

   F They are parallel lines.
   (G) They are intersecting lines.
   H They are perpendicular lines.
   J They are skew lines.

5. How is a pair of perpendicular lines different from a pair of intersecting lines?

   Possible answer: The perpendicular lines must meet to form right angles, while the intersecting lines do not.

6. What kinds of lines are similar to the rails of a railroad?

   parallel lines

LESSON 8-4
Practice B
Classifying Lines

Classify each pair of lines.

1. 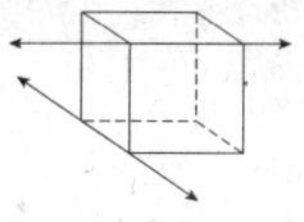

   skew lines

2. 

   intersecting lines

3. 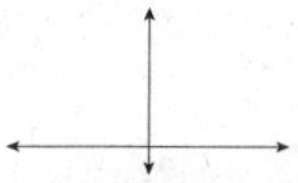

   perpendicular lines

4. 

   parallel lines

Match each description with its correct classification.

5. $\overleftrightarrow{AB}$ and $\overleftrightarrow{EF}$ lie on the same plane and never intersect. B
6. $\overleftrightarrow{AB}$ and $\overleftrightarrow{EF}$ cross each other at one common point. A
7. $\overleftrightarrow{AB}$ and $\overleftrightarrow{EF}$ lie on different planes and are neither parallel nor intersecting. C
8. $\overleftrightarrow{AB}$ and $\overleftrightarrow{EF}$ intersect to form right angles. D

A. $\overleftrightarrow{AB}$ intersects $\overleftrightarrow{EF}$.

B. $\overleftrightarrow{AB} \parallel \overleftrightarrow{EF}$

C. $\overleftrightarrow{AB}$ and $\overleftrightarrow{EF}$ are skew.

D. $\overleftrightarrow{AB} \perp \overleftrightarrow{EF}$

9. Oak Street runs parallel to Elm Street in a flat section of town. Tom tells you to meet him at the intersection of Oak and Elm. Explain why these instructions are impossible to follow.

   Because Oak and Elm are parallel streets on the same plane, they will never intersect.

10. Look around your classroom. Name a pair of parallel lines and a pair of perpendicular lines that you see.

   Answers will vary. Possible parallel lines: the 2 sides of my desk; Possible perpendicular lines: the top and side of the chalkboard

LESSON 8-4 **Practice C**
Classifying Lines

Use the diagram for Exercises 1–3. Possible answers are given.

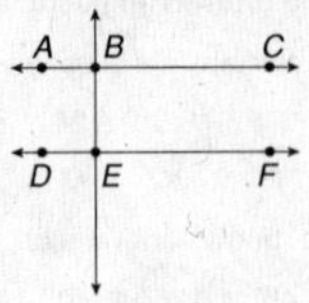

1. Name two pairs of perpendicular lines.

   $\overleftrightarrow{AC}$ and $\overleftrightarrow{BE}$; $\overleftrightarrow{DF}$ and $\overleftrightarrow{BE}$

2. Name a pair of parallel lines.

   $\overleftrightarrow{AC}$ and $\overleftrightarrow{DF}$

3. Name two pairs of intersecting lines.

   $\overleftrightarrow{AC}$ and $\overleftrightarrow{BE}$; $\overleftrightarrow{DF}$ and $\overleftrightarrow{BE}$

The lines in the figure intersect to form a rectangular box. Possible answers

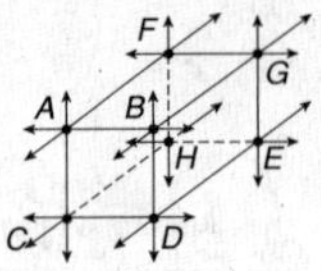

4. Name two pairs of lines that are parallel to $\overleftrightarrow{DE}$.

   $\overleftrightarrow{BG}$ and $\overleftrightarrow{AF}$; $\overleftrightarrow{CH}$ and $\overleftrightarrow{AF}$

5. Name a pair of lines that are skew.

   $\overleftrightarrow{AC}$ and $\overleftrightarrow{EH}$

6. Name all the lines that are perpendicular to $\overleftrightarrow{AB}$.

   $\overleftrightarrow{AC}$, $\overleftrightarrow{AF}$, $\overleftrightarrow{BG}$ and $\overleftrightarrow{BD}$

7. Air traffic controllers must prevent two planes from sharing the same point in the sky at the same time or the planes will collide. What types of flight path lines will always prevent a collision of two planes?

   skew path lines and parallel path lines

8. Line *AD* is perpendicular to line *AB* at point *A*. Line *BC* is perpendicular to line *AB* at point *B*. All the lines lie in the same plane. What is the relationship between line *AD* and line *BC*?

   They are parallel lines.

LESSON 8-4 **Reteach**
Classifying Lines

Some lines have relationships.

Intersecting lines cross each other at one point.

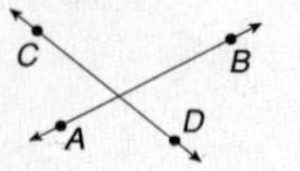

$\overleftrightarrow{AB}$ intersects $\overleftrightarrow{CD}$.

Perpendicular lines intersect to form right angles.

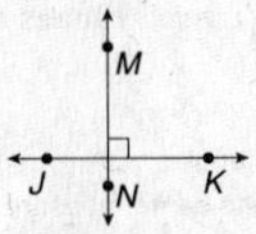

$\overleftrightarrow{JK} \perp \overleftrightarrow{MN}$

Parallel lines lie in the same plane but never intersect.

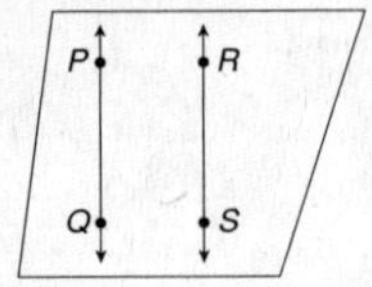

$\overleftrightarrow{PQ} \parallel \overleftrightarrow{RS}$

Skew lines lie in different planes and do not intersect.

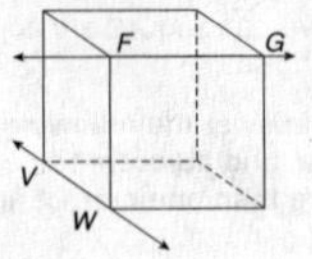

$\overleftrightarrow{FG}$ and $\overleftrightarrow{VW}$ are skew.

Classify each pair of lines.

1.

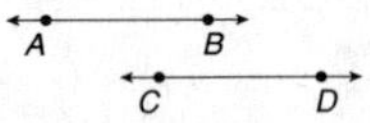

parallel

2.

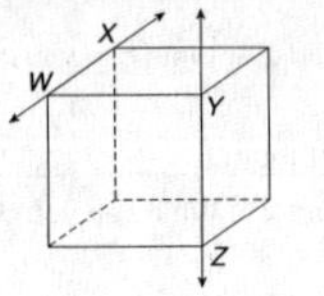

skew

3.

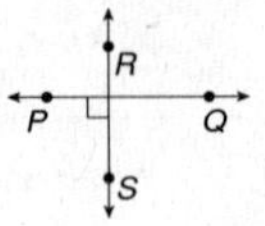

perpendicular

4.

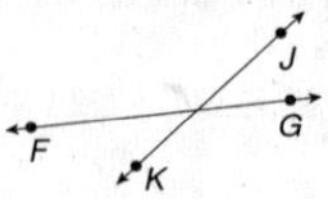

intersecting

LESSON 8-4 **Challenge**
Capital Streets

Washington, D.C., became the capital of the United States in 1800. The city's planners designed wide streets reaching out from two main centers—the White House and the Capitol Building. The President works and lives in the White House, and Congress works in the Capitol.

The map below shows the section of Washington, D.C, where the White House and Capitol are located. Use the following clues to label the streets on the map. Then name all the pairs of perpendicular streets on the map.

1. Madison Drive, Jefferson Drive, and Independence Avenue are parallel streets. Madison Drive is north of Jefferson Drive and Jefferson Drive is north of Independence Avenue.
2. 14th Street and 3rd Street are parallel streets. 14th Street is west of 3rd Street.
3. Pennsylvania Avenue intersects 14th Street near the White House.
4. The National Mall lies between Madison Drive to the north and Jefferson Drive to the south.

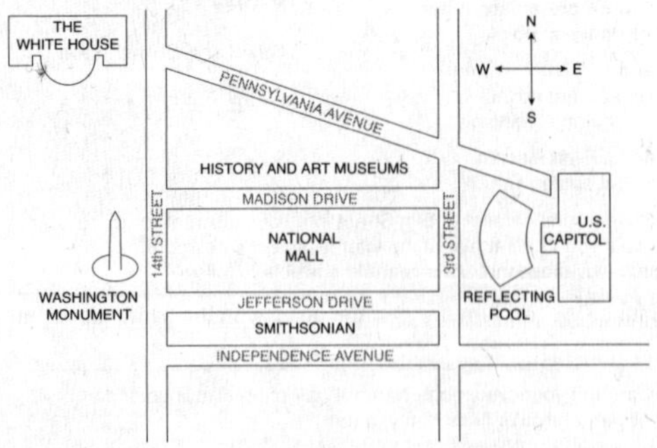

14th ⊥ Madison; 14th ⊥ Jefferson; 14th ⊥ Independence;
3rd ⊥ Madison; 3rd ⊥ Jefferson; 3rd ⊥ Independence

LESSON 8-4 **Problem Solving**
Classifying Lines

Use the map to answer each question.

Western U.S. States

1. The area where the borders of Utah, Colorado, Arizona, and New Mexico meet is sometimes called the Four Corners. What kind of lines are formed where the borders meet?

   perpendicular lines

2. Which borderlines on the map are skew lines?

   none

3. What kinds of lines are suggested by the eastern and western borders of New Mexico?

   parallel lines

Circle the letter of the correct answer.

4. Which three states' borderlines intersect near the Grand Canyon?
   - A Utah, Arizona, and Idaho
   - B Idaho, Arizona, and Oregon
   - (C) Nevada, Utah, and Arizona
   - D Utah, Wyoming, and Idaho

5. Which two western states seem to have congruent borderlines?
   - (F) Colorado and Wyoming
   - G Oregon and Nevada
   - H New Mexico and Nevada
   - J Utah and Idaho

6. Which of the following do not appear to be parallel to the western borderline of Nevada?
   - (A) the western borderline of California
   - B the western borderline of Wyoming
   - C the eastern borderline of Montana
   - D the eastern borderline of Arizona

7. Which of these western states do not have borderlines that intersect near Great Salt Lake?
   - F Utah and Nevada
   - G Utah and Idaho
   - H Utah and Wyoming
   - (J) Utah and Colorado

LESSON 8-4

Reading Strategies
Understanding Symbols

Parallel lines are the same distance apart and never meet.

The lines on a ruled piece of paper are similar to parallel lines.
The symbol II stands for parallel.

Segment *AB* is parallel to segment *CD*.

This can be written: *AB* II *CD*.

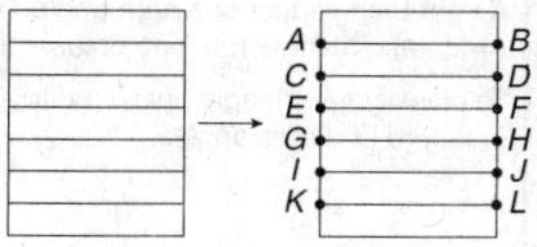

1. Find another pair of parallel line segments in the figure above.

Possible answer: $\overline{GH}$ and $\overline{IJ}$

2. Use the II symbol to write how the line segments are related.

$\overline{GH} \parallel \overline{IJ}$

Perpendicular lines meet to form square corners.

The symbol ⊥ stands for perpendicular. A square corner is used as a symbol to indicate that two lines are perpendicular to each other.

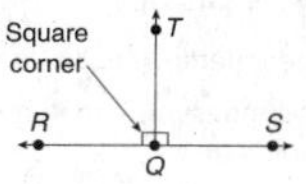

$\overline{TQ} \perp \overline{RS}$

Read: segment *TQ* perpendicular to segment *RS*.

Use this figure to complete Exercises 3–5.

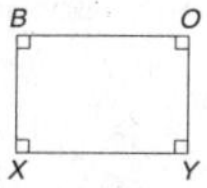

3. Segment *BX* is perpendicular to segment *BO*. Write this with symbols.

$\overline{BX} \perp \overline{BO}$

4. Find two other line segments that are perpendicular to each other.

Possible answer: $\overline{XY}$ and $\overline{OY}$

5. What symbol in the figure tells you that line segments are perpendicular to each other?

the square corner

LESSON 8-4

Puzzles, Twisters & Teasers
What is my Line?

Use the figure on the right. For each box decide whether the statement is true or false. If it is false, black out the box.

$\overline{FG} \perp \overline{HJ}$ A	$\overline{FG} \perp \overline{GJ}$ G	$\overline{HJ} \parallel \overline{FG}$ C
$\overline{FH} \parallel \overline{GJ}$ B	$\overline{FH} \perp \overline{HJ}$ W	$\overline{FH} \parallel \overline{HJ}$ L
$\overline{FG} \parallel \overline{FH}$ R	$\overline{HJ} \perp \overline{GJ}$ S	$\overline{GJ} \parallel \overline{KL}$ D

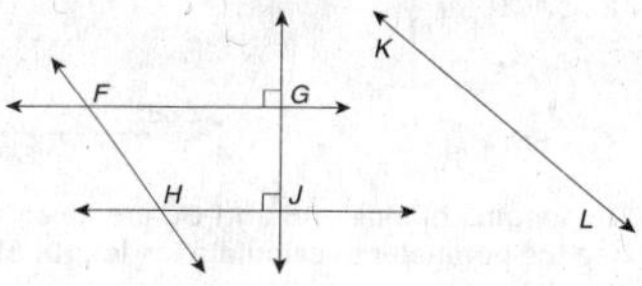

$\overline{AD} \perp \overline{AG}$ E	$\overline{BC}$ & $\overline{CD}$ are skew P	$\overline{DC} \parallel \overline{EF}$ G
$\overline{GF} \parallel \overline{AD}$ I	$\overline{BH} \perp \overline{HE}$ O	$\overline{BH} \parallel \overline{CE}$ M
$\overline{HG} \parallel \overline{AB}$ O	$\overline{HE} \perp \overline{CB}$ F	$\overline{HG}$ & $\overline{BC}$ are skew N

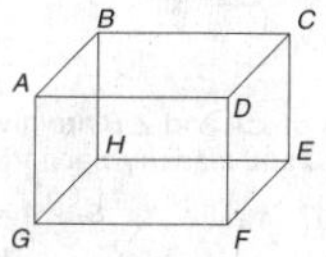

To solve the riddle, re-arrange the letters for the true boxes.

What do hens study in college? eggonomics

LESSON 8-5

Practice A
Triangles

Classify each triangle as acute, obtuse, or right.

1. 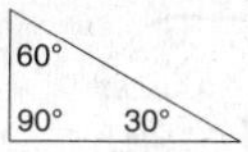

right

2. 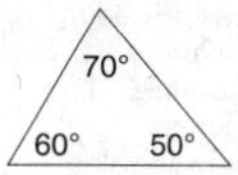

acute

3. 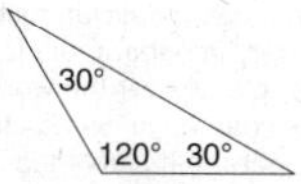

obtuse

Classify each triangle as scalene, isosceles, or equilateral.

4. 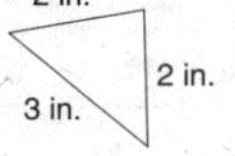

isosceles

5. 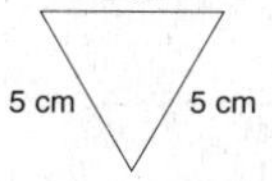

equilateral

6. 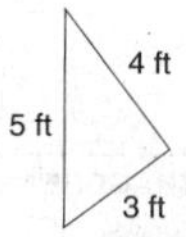

scalene

Find the measure of the unknown angle.

7. 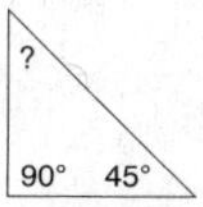

45°

8. 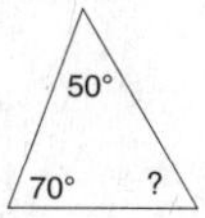

60°

9. 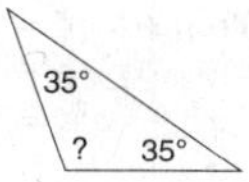

110°

10. One side of an equilateral triangle measures 4 cm. What are the lengths of the other two sides of the triangle?

They both measure 4 cm.

11. Karen says the angles of her triangle measure 90°, 50°, and 60°. Explain why this is impossible.

The sum of those angles is 200°, and the sum of the angles in any triangle is always 180°.

LESSON 8-5

Practice B
Triangles

Use the diagram to find the measure of each indicated angle.

1. ∠*CBD* 90°
2. ∠*DAC* 45°
3. ∠*ABC* 90°
4. ∠*EBA* 90°
5. ∠*ACB* 45°

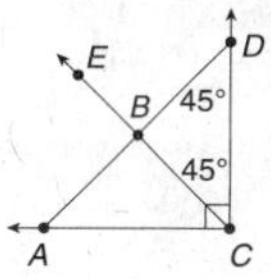

Classify each triangle using the given information.

6. The perimeter of the triangle is 30 in.

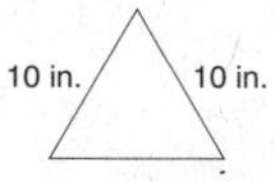

equilateral

7. The perimeter of the triangle is 15 cm.

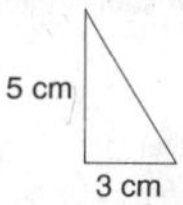

scalene

8. The perimeter of the triangle is 22 ft.

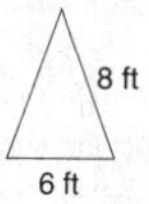

isosceles

9. The angles of a triangular sail measure 90°, 30°, and 60°. Its sides measure approximately 2 feet, 3.5 feet, and 4 feet. Classify the triangular shape of the sail in two different ways.

It is a right triangle; It is a scalene triangle.

10. Two angles in one triangle are congruent to two angles in another triangle. What can you conclude about the third angle in both triangles?

They are congruent.

LESSON 8-5 **Practice C**
Triangles

Use the diagram to find the measure of each indicated angle.

1. ∠DEC **79°**
2. ∠ACB **52°**
3. ∠BCE **117°**
4. ∠EBC **52°**

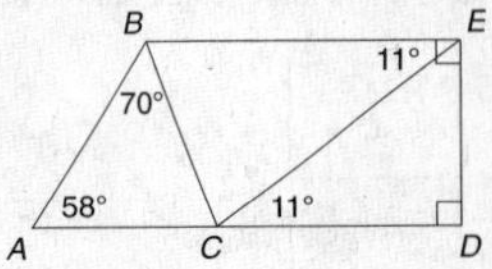

The lengths of sides $\overline{AB}$ and $\overline{BC}$ are given for △ABC. Use the perimeter to calculate the length of side $\overline{AC}$ and classify each triangle.

5. AB = 1.5; BC = 4.8; perimeter = 10
 AC = 3.7; scalene
6. $AB = 2\frac{1}{4}$; $BC = 2\frac{1}{4}$; perimeter = $6\frac{3}{4}$
 $AC = 2\frac{1}{4}$; equilateral
7. AB = 638; BC = 181; perimeter = 1,000
 AC = 181; isosceles

The measures of ∠A and ∠B are given for △ABC. Calculate the measure of ∠C and classify each triangle.

8. ∠A = 22.17°; ∠B = 136.5°
 ∠C = 21.33°; obtuse
9. ∠A = 60°; $∠B = \frac{1}{2} \cdot ∠A$
 ∠C = 90°; right
10. ∠A = 26.5°; ∠B = 3 • ∠A
 ∠C = 74°; acute

11. Starting from the airport, a pilot flew directly north for 112 miles. Then she made a 90° turn and flew east for 84 miles to reach Bear Lake. If she flew in a straight line, in which direction did the pilot fly to get back to the airport? If she flew a total of 336 miles, how far is the airport from Bear Lake?
 She flew southwest to the airport; 140 miles

12. ∠A of △ABC is congruent to ∠G of △FGH. ∠B of △ABC and ∠F of △FGH are supplementary. ∠C measures 29°, and ∠A measures 37°. What is the measure of each angle in △FGH?
 ∠G = 37°; ∠F = 66°; ∠H = 77°

LESSON 8-5 **Reteach**
Triangles

You can classify triangles by the measures of their angles.

An acute triangle has only acute angles.
A right triangle has one right angle.
An obtuse triangle has one obtuse angle.

To classify the triangle below by its angles, you first need to find the measure of all the angles.

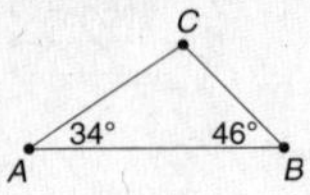

The sum of the angles of a triangle is 180°.

To find the unknown angle, first find the sum of the two known angles.

34 + 46 = 80

Then subtract the sum from 180.

180 − 80 = 100

The difference is the measure of the third angle.

The measures of the angles are 34°, 46°, and 100°. So the triangle is obtuse.

Find each unknown angle. Then classify each triangle.

1. 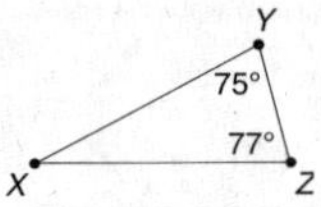

 28°, acute

2. 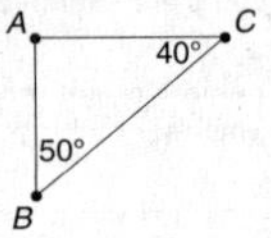

 90°, right

3. 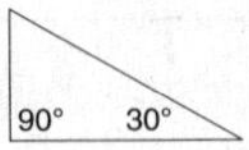

 60°, right

4. 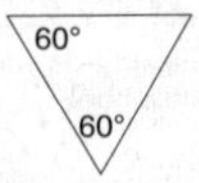

 60°, acute

LESSON 8-5 **Reteach**
Triangles (continued)

You can also classify triangles by the lengths of their sides.

A scalene triangle has no congruent sides.
An isosceles triangle has at least 2 congruent sides.
An equilateral triangle has 3 congruent sides.

The perimeter of the triangle below is 18.6 cm. To classify the triangle below by its sides, you first need to find the length of the unknown side.

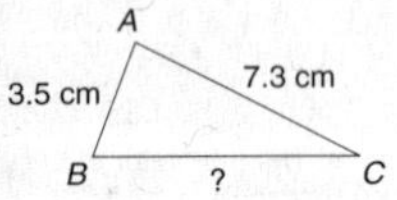

To find the unknown length, first find the sum of the two known lengths.

7.3 + 3.5 = 10.8

Then subtract this sum from the perimeter.

18.6 − 10.8 = 7.8

The length of the third side is 7.8 cm.

The lengths of the sides of the triangle are 3.5 cm, 7.3 cm, and 7.8 cm, so the triangle is scalene.

Find each missing length. Then classify each triangle.

3. 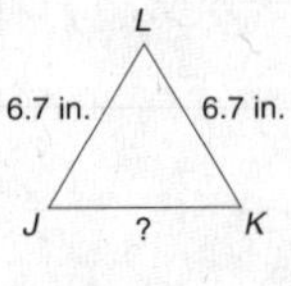

 The perimeter of the triangle is 20.1 inches.
 6.7 inches, equilateral

4. 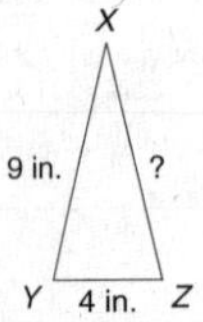

 The perimeter of the triangle is 22 inches.
 9 inches, isosceles

LESSON 8-5 **Challenge**
Square Legs

About 2,400 years ago a Greek philosopher and mathematician named Pythagoras proved a very important rule for triangles. Today people all over the world study and use this rule named in his honor—the Pythagorean Theorem. It states the relationship between the side lengths of a right triangle.

Pythagoras

PYTHAGOREAN THEOREM

In a right triangle, the square of the length of the hypotenuse is equal to the sum of the squares of the lengths of the legs.

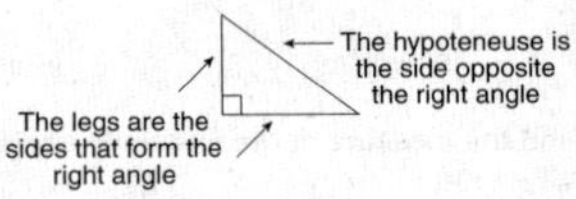

$(\text{leg})^2 + (\text{leg})^2 = (\text{hypotenuse})^2$
↓
$a^2 + b^2 = c^2$
↓
$3^2 + 4^2 = 5^2$
↓
$9 + 16 = 25$

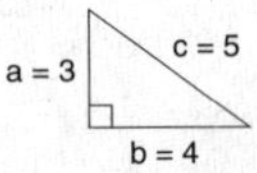

Use the Pythagorean Theorem to find the length of the missing side for each right triangle below.

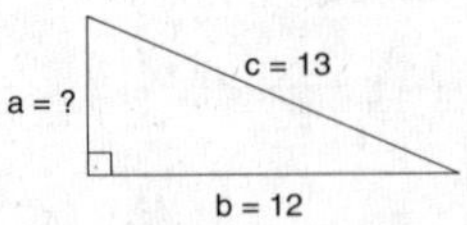

a = 5

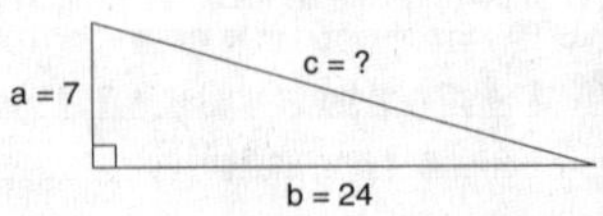

c = 25

LESSON 8-5

Problem Solving
Triangles

Use the triangle diagram to answer each question.

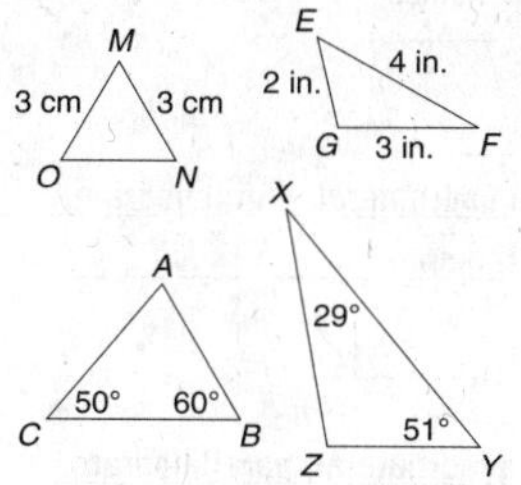

1. Classify triangle *ABC*. What is the measure of the missing angle?

   $\triangle ABC$ is an acute triangle, 70°

2. Classify triangle *XYZ*. What is the measure of the missing angle?

   $\triangle XYZ$ is an obtuse triangle, 100°

3. If triangle *MNO* is an equilateral triangle, what is the measure of the missing side?

   ON = 3 cm

Circle the letter of the correct answer.

4. What is the complement of ∠*XYZ*?
   - (A) 39° (circled)
   - B 51°
   - C 129°
   - D 309°

5. Classify triangle *EFG*.
   - (F) scalene triangle (circled)
   - G isosceles triangle
   - H equilateral triangle
   - J right triangle

6. Which of the following statements is always true?
   - A A right triangle is a scalene triangle.
   - (B) An equilateral triangle is an isosceles triangle. (circled)
   - C An isosceles triangle is an obtuse triangle.
   - D A right triangle is an acute triangle.

7. Which of the following is not true of all right triangles?
   - F The sum of the measures of the angles is 180°.
   - (G) Two of its angles are supplementary angles. (circled)
   - H At least two of its angles are acute.
   - J The side with the greatest length is opposite the right angle.

LESSON 8-5

Reading Strategies
Graphic Organizer

A triangle can be classified by the measurement of one or more of its angles. The length of its sides can also classify a triangle.

This chart will help you compare triangles by their angles and by their sides.

Triangles Named by Angle Measure	**Triangles Named by Length of Sides**
Acute Triangle All three angles measure less than 90°.	**Equilateral Triangle** All the sides are equal.
Right Triangle One angle measures 90°.	**Isosceles Triangle** Two sides are equal.
Obtuse Triangle One angle measures more than 90°.	**Scalene Triangle** None of the sides are equal.

Write two of the following terms to describe each triangle: ***acute, right, obtuse, equilateral, isosceles, scalene.***

1.

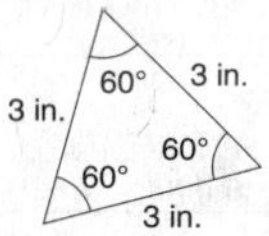

acute, equilateral

2.

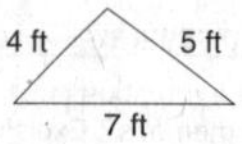

obtuse, scalene

LESSON 8-5

Puzzles, Twisters & Teasers
Find the Measure

Find the missing angle measure in each triangle. When you have completed the problem, look at the blank lines at the bottom of the page. Notice if any of the angle degrees of the letters match the degrees under the blank lines. If they do, write the letter of the angle on the line. The resulting message will be the answer to the riddle.

1.

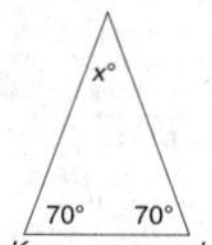

x = 40°

2.

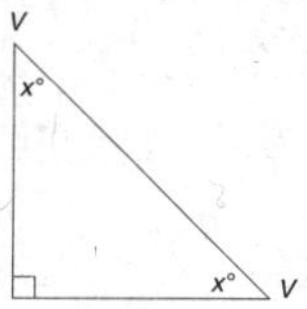

x = 45°

3.

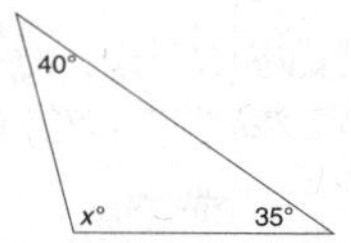

x = 105°

4.

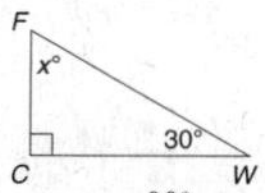

x = 60°

5.

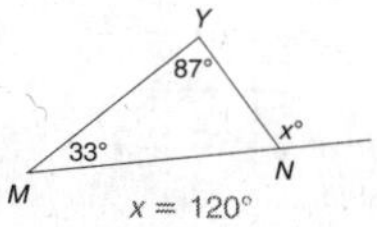

x = 120°

6.

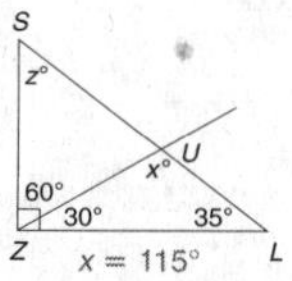

x = 115°
z = 55°

7.

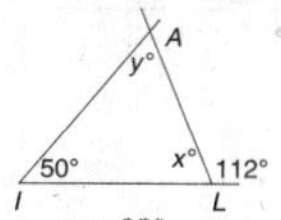

x = 68°
y = 62°

Why do you like studying about volcanoes?

I	T '	S	A	L	A	V	A	F	U	N
50°	105°	55°	62°	68°	62°	45°	62°	60°	115°	120°

LESSON 8-6

Practice A
Quadrilaterals

Choose the correct answer.

1. Which name is most descriptive of the figure shown below?

   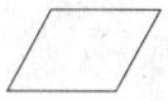

   - A quadrilateral
   - B parallelogram
   - C square
   - (D) rhombus (circled)

2. Which name is most descriptive of the figure shown below?

   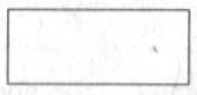

   - F parallelogram
   - G square
   - (H) rectangle (circled)
   - J quadrilateral

3. Which of the following statements is not always true?
   - A A quadrilateral has 4 sides.
   - B A quadrilateral has 4 angles.
   - C A quadrilateral has straight sides.
   - (D) A quadrilateral has right angles. (circled)

4. Which of the following statements is always true?
   - F A trapezoid is also a parallelogram.
   - (G) A square is also a rhombus. (circled)
   - H A parallelogram is also a rectangle.
   - J A rectangle is also a square.

5. Carol says this poster is a parallelogram. Anita says it is a rectangle. Who is right? Explain.

   They both are right. A rectangle is also a parallelogram because its opposite sides are parallel and congruent, and its opposite angles are congruent.

6. Joel hung two square mirrored tiles side-by-side in his bathroom. What is the best description for the geometric figure the two mirrors formed together?

   a rectangle

LESSON 8-6
Practice B
Quadrilaterals

Give the most descriptive name for each figure.

1. 
rhombus

2. 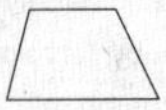
trapezoid

3. 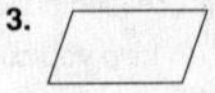
parallelogram

4. 
quadrilateral

5. 
square

6. 
rectangle

Complete each statement.

7. All rectangles are also parallelograms.

8. A rhombus is sometimes a square.

9. All trapezoids are also quadrilaterals.

10. A quadrilateral is any plane figure with four straight sides and four angles.

11. A quadrilateral with two sets of parallel lines, but does not have 90° angles is called a parallelogram.

12. Devon made a table top in the shape of a quadrilateral. All of its angles measure 90°. What could the shape of Devon's table top be?

a rectangle or a square

13. The perimeter of a rhombus is 64 inches. What is the length of each side of the rhombus? Explain.

16 inches, All four sides of a rhombus are congruent, and $64 \div 4 = 16$.

14. Explain why a trapezoid is a quadrilateral, but a quadrilateral is not always a trapezoid.

A trapezoid has four sides, always making it a quadrilateral. A quadrilateral can,but not always, have two parallel sides, thus a quatrilateral is not always a trapezoid.

LESSON 8-6
Practice C
Quadrilaterals

Give all the possible names for each figure. Circle the most exact name.

1. 
quadrilateral, parallelogram, (rhombus)

2. 
quadrilateral, (trapezoid)

3. 
quadrilateral, parallelogram, (rectangle)

4. 
quadrilateral, parallelogram, (square), rectangle, rhombus

Draw each quadrilateral as it is described. If it is not possible, explain.

5. a parallelogram that is also a trapezoid

impossible because a trapezoid has only two parallel sides

6. a rhombus that is not a parallelogram

impossible because a rhombus is always a parallelogram

7. Draw an example of a quadrilateral that is not a parallelogram or a trapezoid.

Students should draw a four-sided figure with no parallel sides.

8. All the sides of a quadrilateral measure 25 cm. Explain why this is not enough information to classify what kind of quadrilateral it is.

It could either be a square or a rhombus. I would need to know the angle measurements to further classify it.

LESSON 8-6
Reteach
Quadrilaterals

A quadrilateral is a plane figure with four sides and four angles.

There are special types of quadrilaterals.

parallelogram rectangle rhombus square trapezoid

A parallelogram has opposite sides that are parallel and congruent. Opposite angles are also congruent.

A rectangle is a parallelogram that has four right angles.

A rhombus is a parallelogram with four congruent sides.

A square is a rectangle with four congruent sides.

A trapezoid is a quadrilateral with exactly two parallel sides.

The best description for the figure below is rectangle because it is a parallelogram with four right angles.

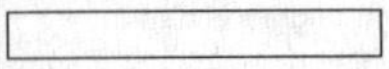

Give the best name for each figure.

1. 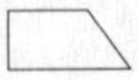
trapezoid

2. 
square

3. 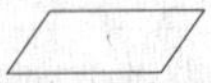
parallelogram

4. 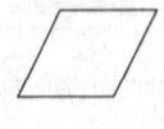
rhombus

5. 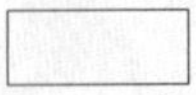
rectangle

6. 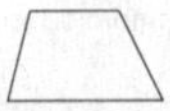
trapezoid

LESSON 8-6
Challenge
What's the Point?

For each grid below, name the ordered pair of the missing fourth point that could be used to complete the given quadrilateral.

1. 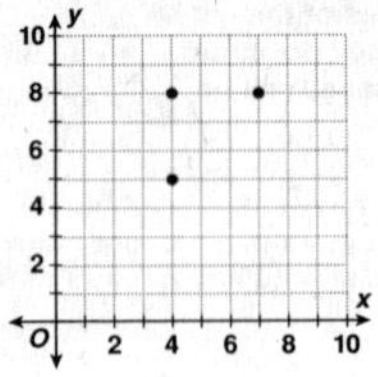
square
(7, 5)

2. 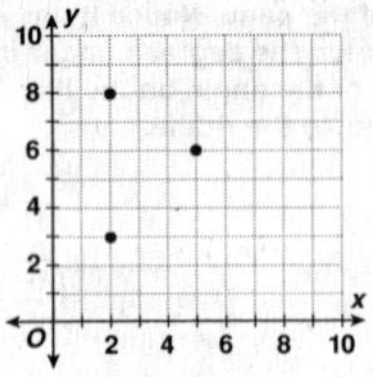
trapezoid
Possible answer: (5, 5)

3. 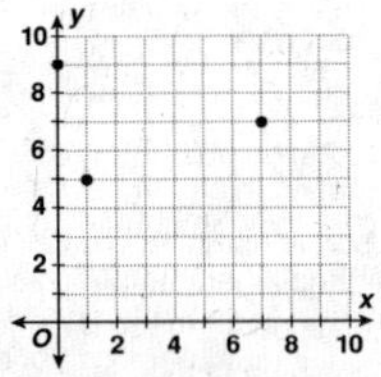
quadrilateral
Possible answer: (3, 9)

4. 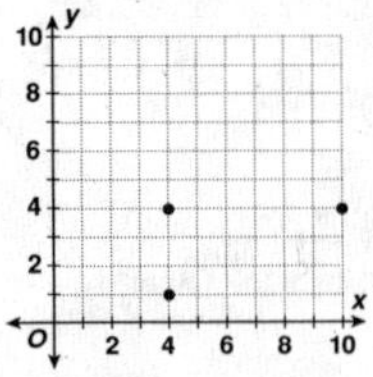
rectangle
(10, 1)

5. 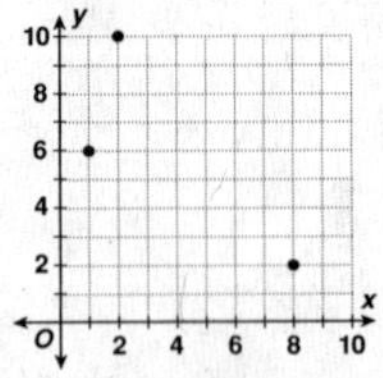
parallelogram
(9, 6)

6. 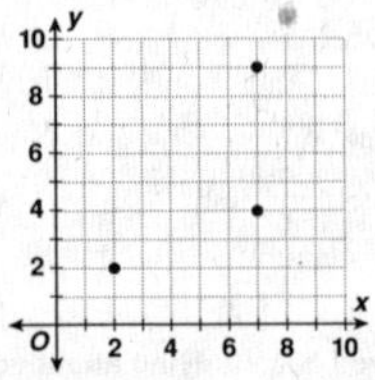
rhombus
(2, 7)

LESSON 8-6
Problem Solving
Quadrilaterals

Write the correct answer.

1. Fill in this Venn diagram using the terms quadrilaterals, squares, rectangles, rhombuses, parallelograms, and trapezoids.

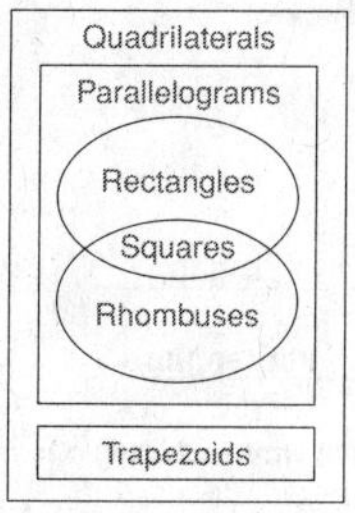

2. Part of this quadrilateral is hidden. What could it possibly be?

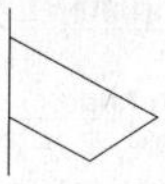

a trapezoid, a parallelogram

3. How could you make a trapezoid from a rectangle using only one cut?

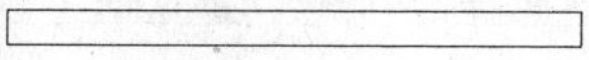

Cut a right triangle from one side

4. An engineer wants to build a building with a parallelogram base. He wants the four corners to be right angles and the four sides congruent. What type of base does the engineer want?

square

Circle the letter of the correct answer.

5. Each side of a quadrilateral-shaped picture frame has the same length. Which of the following is not a possible shape for the frame?
   - A a rhombus
   - B a square
   - (C) a trapezoid
   - D a parallelogram

6. The total length of the four sides of the picture frame from Exercise 5 is 4 feet, 8 inches. What is the length of each of its sides?
   - (F) 14 inches
   - G 1 foot, 3 inches
   - H 12 inches
   - J 2 inches

LESSON 8-6
Reading Strategies
Focus on Vocabulary

The chart below will help you understand the different kinds of quadrilaterals.

A **quadrilateral** is a plane figure with four sides and four angles.

A **trapezoid** is a quadrilateral.
- It has exactly two parallel sides.
- It may have two right angles.

A **parallelogram** is a quadrilateral.
- Both sets of opposite sides are parallel and congruent.
- Opposite angles are congruent.

A **rectangle** is a quadrilateral.
- It is also a parallelogram with four right angles.

A **rhombus** is a quadrilateral.
- It is also a parallelogram with four congruent sides.

A **square** is a quadrilateral.
- It is also a parallelogram.
- It is also a rectangle with four congruent sides.
- It is also a rhombus with four right angles.

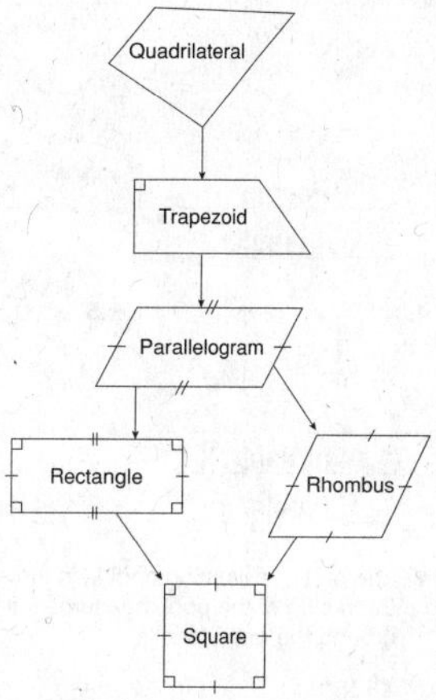

Use the chart to help you answer each question.

1. What is the name of a parallelogram with all sides equal? rhombus
2. What is the name of a rectangle with all sides equal? square
3. What is the name of a quadrilateral with only one pair of parallel sides? trapezoid
4. Name the quadrilaterals that are also parallelograms.

rectangle, rhombus, and square

LESSON 8-6
Puzzles, Twisters & Teasers
Cafeteria Food

Decide if each statement is true or false, and circle the answer.

Answer the riddle by rearranging the letters next to your circled answers.

1. All squares are quadrilaterals.	(True S)	False L
2. A trapezoid is a quadrilateral with exactly two parallel sides.	(True R)	False C
3. Some squares are not rectangles.	True O	(False M)
4. A rhombus is a parallelogram with two congruent sides.	True J	(False O)
5. Some quadrilaterals are not squares.	(True M)	False B
6. Opposite angles of a parallelogram are not always congruent.	True N	(False U)
7. A rectangle is a parallelogram.	(True H)	False K
8. A triangle is a quadrilateral.	True L	(False O)

What do you call the place where school lunches are made? mushroom

LESSON 8-7
Practice A
Polygons

Name each polygon.

1. 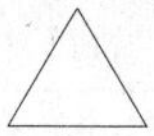 triangle
2.  hexagon
3. 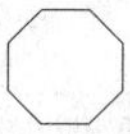 octagon
4.  quadrilateral
5. 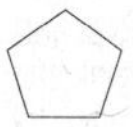 pentagon
6. 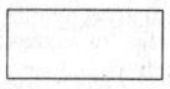 quadrilateral

Tell whether each polygon appears to be *regular* or *not regular*.

7. 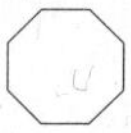 regular
8. 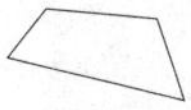 not regular
9. 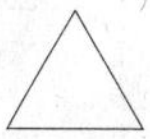 regular

10. What is another name for a regular triangle?

an equilateral triangle

11. What is another name for a regualr quadrilateral with four right angles?

a square

12. One side of a regular pentagon measures 4 inches. What is the total length of the sides of that pentagon?

20 inches

13. The total length of the sides of a regular octagon is 24 inches. What is the length of each side?

3 inches

LESSON 8-7
Practice B
Polygons

Name each polygon and tell whether it appears to be *regular* or *not regular*.

1. 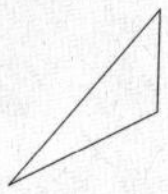
triangle; not regular

2. 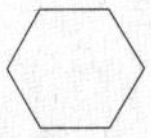
hexagon; regular

3. 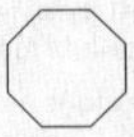
octagon; regular

4. 
quadrilateral; regular

5. 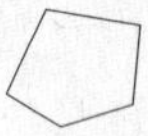
pentagon; not regular

6. 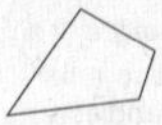
quadrilateral; not regular

7. The public swimming pool is in the shape of a regular hexagon. Each side of the pool measures 5 feet. What is the distance around the entire pool?

30 feet

8. In the space below, draw a regular quadrilateral. Now draw one diagonal of that quadrilateral. Describe the two polygons that are formed.

They are both triangles.

LESSON 8-7
Practice C
Polygons

Name each polygon and tell whether it appears to be *regular* or *not regular*.

1. 
hexagon; not regular

2. 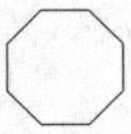
octagon; regular

3. 
quadrilateral; not regular

Explain why each shape is *not* a polygon.

4. 
It is not a closed figure.

5. 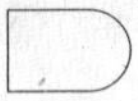
One of its sides is not a line segment.

6. 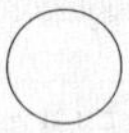
It does not have any line segments.

7. The perimeter of a flower bed designed in the shape of a regular hexagon is $28\frac{1}{2}$ feet. What is the length of each side of the flower bed?

$4\frac{3}{4}$ feet

8. How many diagonals can you draw from one vertex of a regular hexagon? What is the least number of triangles needed to fill a regular hexagon?

3 diagonals, 4 triangles

9. What is the total number of diagonals that can be drawn in a regular pentagon? Describe the figure formed by the diagonals.

5 diagonals, a 5-cornered star

LESSON 8-7
Reteach
Polygons

A polygon is a closed plane figure formed by three or more line segments.

Polygons are named by the number of their sides and angles.

Sides and Angles	3	4	5	6	8
Polygon	triangle	quadrilateral	pentagon	hexagon	octagon

A regular polygon has all congruent sides and angles.

Each angle is 90 degrees.
Each side is 3 centimeters long.

If the angles and the sides are not all congruent, then the polygon is not regular.

Name each polygon and tell whether it is regular or not regular.

1. 
quadrilateral, regular

2. 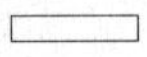
quadrilateral, not regular

To find the angle measures of a regular hexagon, first divide the polygon into triangles by drawing connecting segments from one vertex to all of its nonadjacent vertices.

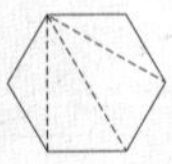

There are 4 triangles.

Next, multiply the number of triangles by 180.

$4 \cdot 180 = 720$ The sum of the angle measures of a triangle is 180°.

Divide the product by 6, the number of sides, to find the measure of each angle.

$\frac{720}{6} = 120$ All of the angles of a regular polygon are congruent.

Each angle is 120°.

Find the angle measure of each regular polygon.

3. regular quadrilateral 90°

4. regular pentagon 108°

LESSON 8-7
Challenge
Sign Language

Traffic signs are symbols used to convey information. In the United States, certain shapes are used only for certain traffic signs. As a result, you can often understand the message of a traffic sign simply by classifying its geometric shape.

Tell whether each traffic sign below is a polygon. If it is, name the polygon that best describes its shape, and tell whether it appears to be *regular* or *not regular*.

1. 

yes, octagon, regular

2. 

not a polygon

3. 

yes, triangle, regular

4. 
yes, pentagon, not regular

5. 

yes, quadrilateral, not regular

6. 

not a polygon

7. 

yes, trapezoid, not regular

8. 
yes, quadrilateral, regular

9. 

yes, triangle, not regular

LESSON 8-7
Problem Solving
Polygons

Write the correct answer.

1. Name each polygon in this figure.

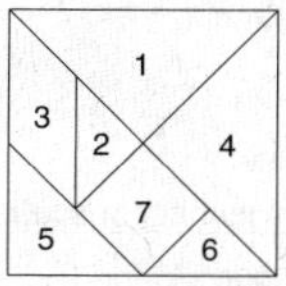

1, 2, 4, 5, and 6 are triangles; 3 is a parallelogram; 7 is a square (some students may name the entire figure a square)

2. How could you use the sum of the angles inside a triangle to find the sum of the angles inside a heptagon?

Five triangles are needed to fill a heptagon; 5 × 180° = 900°. So, the sum of the angles inside a heptagon is 900°.

3. How could you use the sum of the angles inside a triangle to find the sum of the angles inside a decagon?

Eight triangles are needed to fill a decagon; 8 × 180° = 1,440°. So, the sum of the angles inside a decagon is 1,440°.

4. In the space below, draw a rectangle and a parallelogram with side lengths congruent to the rectangle's. Now draw the diagonals for each of those polygons. What new polygons are formed by the diagonals in each quadrilateral?

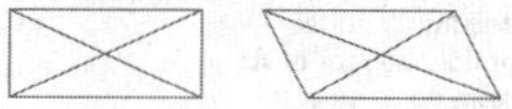

triangles

5. In Exercise 4, what is true of the diagonals in the rectangle that isn't true of the diagonals of the parallelogram?

The diagonals in the rectangle are congruent.

Circle the letter of the correct answer.

6. The perimeter of a regular hexagon is $13\frac{1}{2}$ inches. What is the length of each side?

A $2\frac{7}{10}$ inches
C $3\frac{3}{4}$ inches
(B) $2\frac{1}{4}$ inches
D $1\frac{11}{16}$ inches

7. Which of the following statements is sometimes false?

(F) A plane figure is a polygon.
G Each side of a polygon intersects exactly two other sides.
H A polygon is a closed figure.
J A polygon has straight sides.

LESSON 8-7
Reading Strategies
Focus on Vocabulary

A **polygon** is a closed, plane figure formed by three or more line segments.

In a **regular polygon,** the sides are of equal length and all angle measures are the same.

Regular Polygon **Not Regular Polygon**

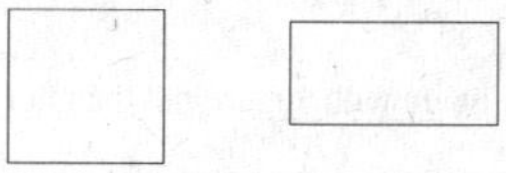

The prefix of a polygon is a clue that can help you understand the number of sides and angles it has.

- *Tri-* means "three" → A **tri**ple play means three outs all at once.
  → A **tri**angle has three sides and three angles.
- *Quad-* means "four" → **Quad**ruplets are four babies born at the same time.
  → A **quad**rilateral has four sides and four angles.

This table helps you organize polygons by their sides and angles.

Polygon	Number of Sides	Number of Angles
Triangle	3 sides	3 angles
Quadrilateral	4 sides	4 angles
Pentagon	5 sides	5 angles
Hexagon	6 sides	6 angles
Octagon	8 sides	8 angles

Write the name of each polygon. Also write whether the polygon is *regular* or *not regular.*

1. 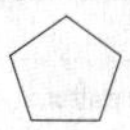

pentagon, regular

2. 

pentagon, not regular

3. 

quadrilateral, not regular

4. 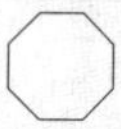

octagon, regular

LESSON 8-7
Puzzles, Twisters & Teasers
Seeing is Believing

There are three correct answers for each problem. Circle the single incorrect statement. The letters next to the incorrect statement will provide the solution to the riddle when they are placed on the blank lines, which match the numbers below.

1. 

T has 6 angles
(F) is an octagon
N is an irregular polygon
A has 6 sides

2. 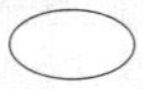

U has no sides
E is an oval
A is not a quadrilateral
(Y) is a polygon

3. 

L has 3 sides
K is an isosceles triangle
(P) is not a polygon
D sum of interior angles equals 180°

4. 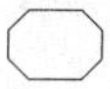

O is an octagon
(A) sum of the interior angles is 720°
G can be divided into 6 triangles
S has 8 angles

5. 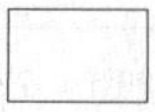

P has 4 right angles
(L) is not a quadrilateral
H has 4 sides
W sum of interior angles equals 360°

6. 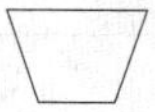

K is a polygon
(S) has 4 congruent sides
N is a trapezoid
T is a quadrilateral

7. 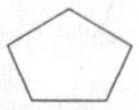

(O) sum of interior angles equal 560°
U has 5 sides
E can be divided into 3 triangles
A has 5 angles

8. 

J is a square
M is a regular polygon
F sum of the interior angles is 360°
(C) sides are not congruent

What kind of glasses do spies use?

S	P	Y	F	O	C	A	L	S
6	3	2	1	7	8	4	5	6

LESSON 8-8
Practice A
Geometric Patterns

Identify a possible pattern. Use the pattern to draw the next figure.

1. 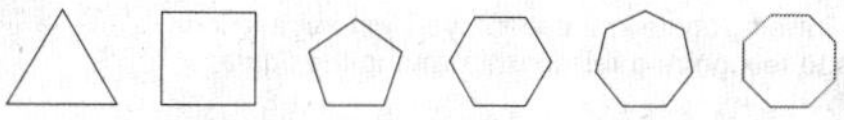

Pattern: Each polygon has one side more than the polygon before it.

2. 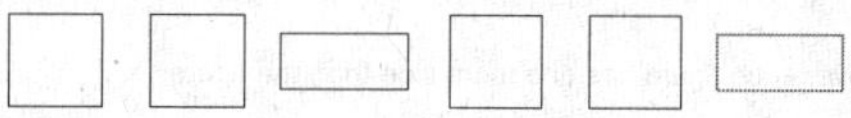

Pattern: Two squares are followed by 1 rectangle

3. 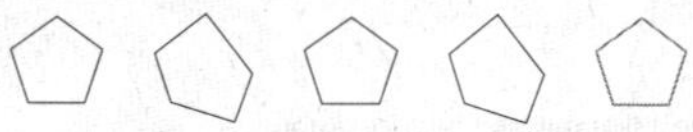

Pattern: Repeating groups of 1 regular pentagon followed by 1 irregular pentagon

4. 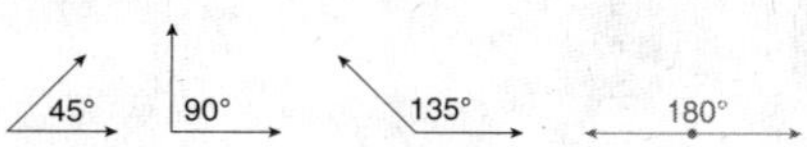

Pattern: Each angle measure increases by 45°.

5. Look around your classroom. Describe a geometric pattern you see.

Possible patterns to describe: the arrangement of desks, ceiling or floor tiles, bulletin board borders

6. Use squares to create a geometric pattern. Describe your pattern.

Accept all geometric patterns using squares. Students' descriptions should match their patterns.

LESSON 8-8

Practice B
Geometric Patterns

Identify a possible pattern. Use the pattern to draw the next figure.

1.

Pattern: A bottom row is added to each figure with 1 more dot than the row above it.

2.

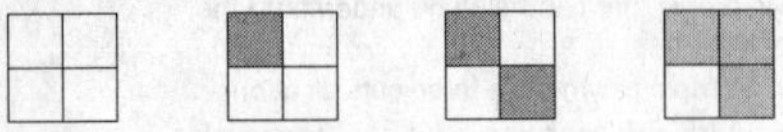

Pattern: One more square is shaded than the figure before it.

3.

Pattern: Repeating groups of 1 right triangle, 1 equilateral triangle, and 1 scalene triangle

4.

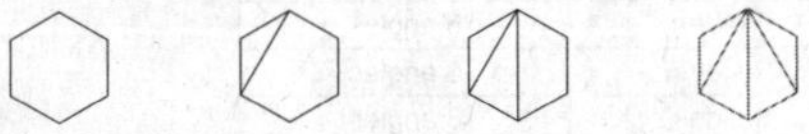

Pattern: One more diagonal is drawn from the same vertex in each hexagon in the pattern.

5. Use triangles to create a geometric pattern. Describe your pattern.

Accept all geometric patterns using triangles. Students' descriptions should match their patterns.

LESSON 8-8

Practice C
Geometric Patterns

Identify a possible pattern. Use the pattern to draw the next figure.

1.

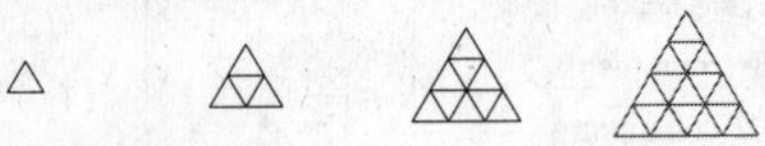

Pattern: The number of equilateral triangles in each figure is the next square number.

2.

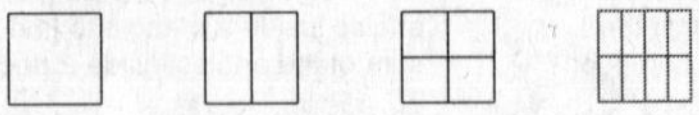

Pattern: Each square is divided into 2 times as many equal parts as the square before it.

3.

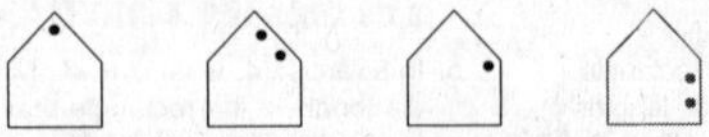

Pattern: The dots move clockwise around the inside of the pentagon. They alternate one dot in a vertex, and 2 dots along a side.

4. Three squares, **A**, **B**, and **C**, are each divided into 9 equal squares. The top left square and the bottom right square of **A** are shaded. The middle top and middle bottom squares are shaded in **B**. In **C**, the top right and the bottom left squares are shaded. Describe **D**, the next square in the pattern.

It will be divided into 9 equal squares, with the middle squares on each side shaded.

5. Use pentagons to create a geometric pattern. Describe your pattern.

Accept all geometric patterns using pentagons. Students' descriptions should match their patterns.

LESSON 8-8

Reteach
Geometric Patterns

Sometimes, patterns involve geometric shapes and figures. The pattern can be based on shape, color, size, position, or quantity.

Before you draw a missing figure in a pattern, you first have to look at the other figures to recognize a relationship among the figures.

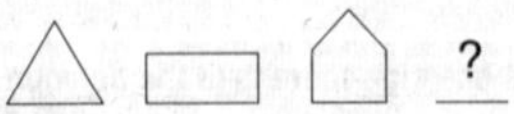

In the pattern above, each figure has one more side than the figure that proceeds it.

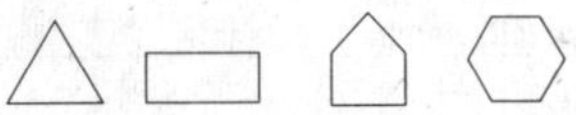

To continue the pattern, the next figure would be a hexagon.

Identify a pattern for each. Then draw the missing figure.

1.

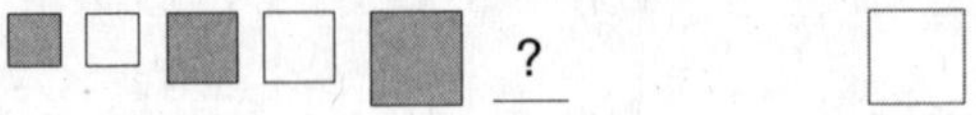

2.

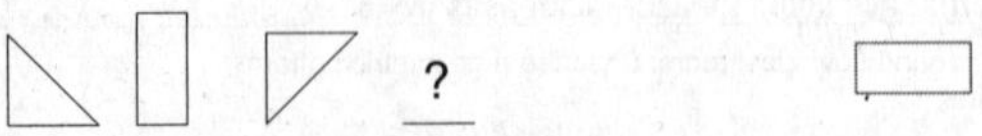

3.

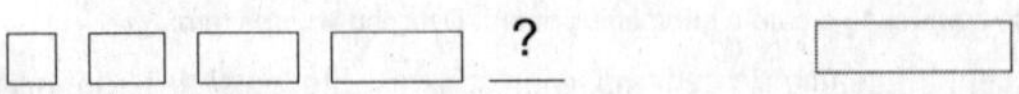

LESSON 8-8

Challenge
Polygon Patterns

Look for patterns to complete this chart and discover two rules for polygons.

Regular Polygon	Number of Sides, *n*	Number of Triangles to Fill	Sum of the Interior Angles
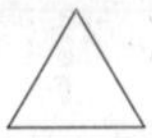 Triangle	$n = 3$	1	1(180°) = 180°
 Quadrilateral	$n = 4$	2	2(180°) = 360°
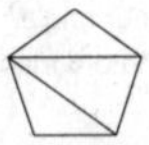 Pentagon	$n = 5$	3	3(180°) = 540°
 Hexagon	$n = 6$	4	4(180°) = 720°

1. Use the patterns to write an expression for the number of triangles needed to fill any regular polygon.

$n - 2$

2. Use the patterns to write an expression for the sum of all the angles inside any regular polygon.

$(n - 2)180°$

3. Using these expressions, how many triangles are needed to fill any regular octagon? What is the sum of a regular octagon's interior angles?

6; 1,080°

LESSON 8-8

Problem Solving

Geometric Patterns

Complete this chart and look for patterns. Then answer the questions. Possible answers are given for 1–4.

	Number of Points on the Line	Draw and Label the Line and Points	Number of Different Line Segments in the Line
	2	A B	1
1.	3	A B C	3
2.	4	A B C D	6
3.	5	A B C D E	10
4.	6	A B C D E F	15

Circle the letter of the correct answer.

5. If n = the number of points on a line, which of the following expressions shows the number of different line segments on that line?
   - A $2n - 3$
   - (B) $(n^2 - n) \div 2$
   - C $(n \div 2) \cdot 5$
   - D $10n \div 2$

6. Using the pattern in the table and your answer to Exercise 5, how many different line segments will be on a line if there are 10 points on the line?
   - F 17 line segments
   - G 25 line segments
   - (H) 45 line segments
   - J 50 line segments

LESSON 8-8

Reading Strategies

Analyze Information

You can examine geometric figures and notice a pattern that forms from one figure to the next.

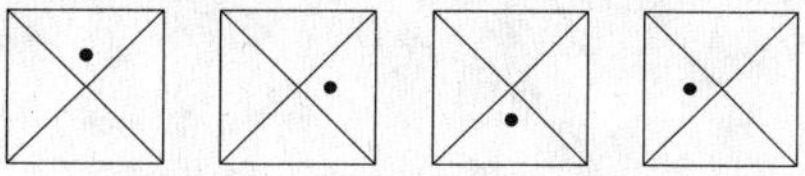

- The black dot is in the **top** section of the figure.
- The black dot is in the **right** section of the figure.
- The black dot is in the **bottom** section of the figure.
- The black dot is in the **left** section of the figure.

1. If the pattern continues, where will the dot be in the next figure?

   in the top section of the figure

2. Draw the dot where it should be to continue this pattern.

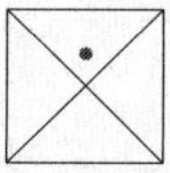

Examine these figures and complete Exercises 3 and 4.

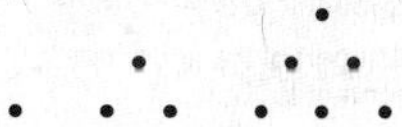

3. What do you notice about the number of dots in the bottom row of each figure?

   A dot is added to the bottom row with each new figure.

4. Draw the next figure to continue the pattern.

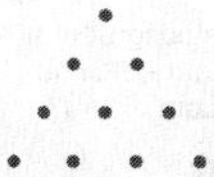

LESSON 8-8

Puzzles, Twisters & Teasers

What's Next?

Draw lines to match the next pattern in the sequence. Not all patterns will match. Once you have all the lines drawn go back to number one and put the letter that is next to the answer in the puzzle answer.

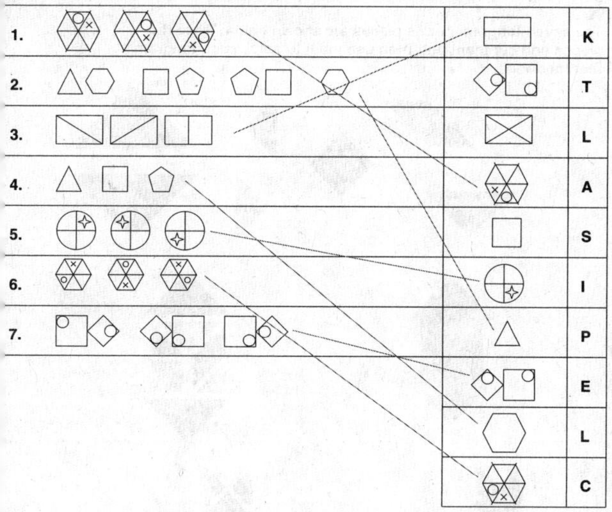

What do you call a cucumber in a sour mood?

A		P	I	C	K	L	E
1		2	5	6	3	4	7

LESSON 8-9

Practice A

Congruence

Decide whether the figures in each pair are congruent. If not, explain.

1. 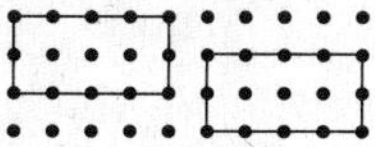

   congruent

2. 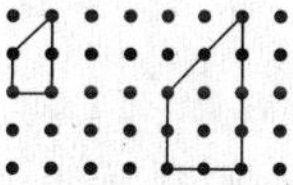

   not congruent; they have different sizes.

3. 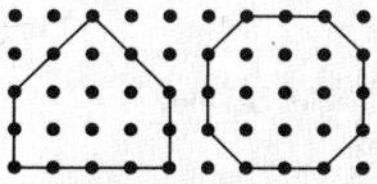

   not congruent; they have different shapes.

4. 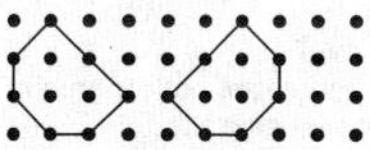

   congruent

Circle the two figures in each row that are congruent.

5. 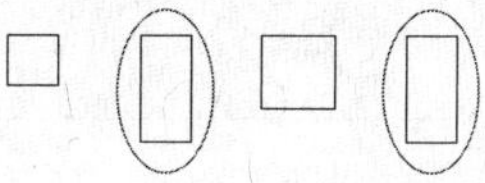

6. 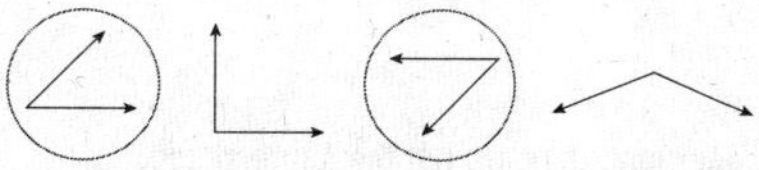

7. Use a ruler to draw two congruent line segments.

   Students should draw two line segments with the same length.

8. If you know that two figures are congruent, what do you know about them?

   They have the same size and shape.

LESSON 8-9
Practice B
Congruence

Decide whether the figures in each pair are congruent. If not, explain.

1. 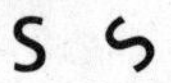

congruent

2. 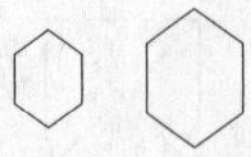

not congruent; they have different sizes

3. 

not congruent; they have different lengths

4. 

congruent

Use the diagram for Exercises 5–7.

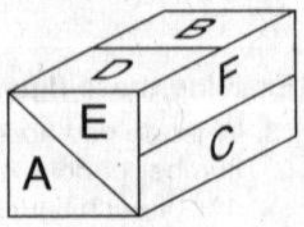

5. Which part of the figure is congruent to A? E
6. Which part of the figure is congruent to D? B
7. Which part of the figure is congruent to F? C
8. Name two parts of your body that appear to be congruent.

Possible answers: eyes, ears, arms, legs, feet, hands

9. Square *ABCD* is congruent to square *FGHJ*. The total length of the sides of square *ABCD* is 12 meters. What is the length of each side of square *FGHJ*?

3 meters

LESSON 8-9
Practice C
Congruence

Decide whether the figures in each pair are congruent. If not, explain.

1. 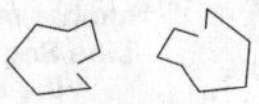

congruent

2. 

not congruent; they have different sizes

Use the dot paper to draw a congruent figure next to the given figure.

3. 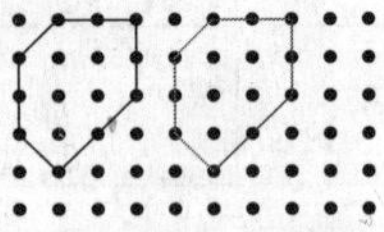

4. 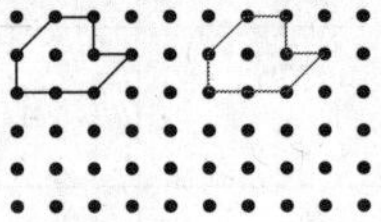

5. Draw a line inside this triangle to create two congruent figures. Describe what type of line you drew and classify the congruent figures you created.

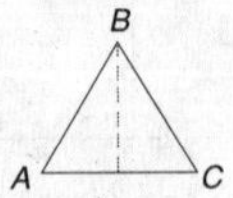

Possible answer: I drew a line perpendicular to side *AC* from vertex *B*; I created two congruent right triangles.

6. Name all the congruent parts of this glass box.

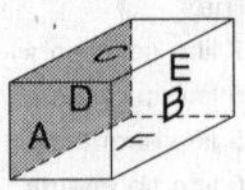

A is congruent to E; B, C, D, and F are congruent.

LESSON 8-9
Reteach
Congruence

Two figures are congruent if they have the same size and shape.

Look at the two figures.

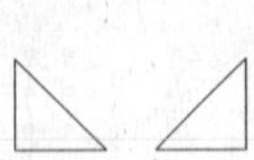

They are congruent because they are both triangles and they are the same size.

Now look at these two figures.

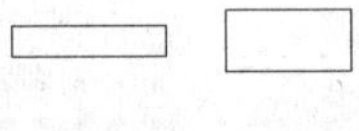

These figures are not congruent. They are both rectangles but they are not the same size.

Decide whether the figures in each pair are congruent. If not, explain.

1. 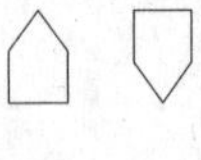

The figures are congruent.

2. 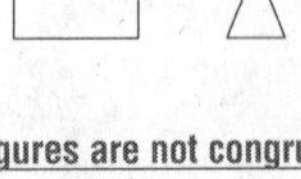

The figures are not congruent; they are not the same shape.

3. 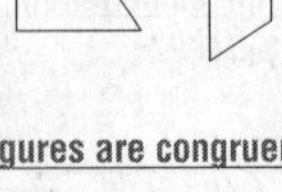

The figures are congruent.

LESSON 8-9
Challenge
Tangram

Tangram is an ancient Chinese puzzle made by cutting a square into 7 pieces—2 congruent small triangles, 2 congruent large triangles, 1 medium triangle, a square, and a parallelogram. To solve a Tangram, you need to rearrange those seven pieces to exactly reproduce a given image.

The seven Tangram puzzle pieces are shown below. Trace the pieces and cut them out. Then use them to solve each Tangram that follows.

1. 

2. 

LESSON 8-9

Problem Solving

Congruence

Write the correct answer.

1. Similar figures have the same shape but may have different sizes. How are similar figures different from congruent figures?

   Congruent figures must have the same shape and size.

2. Pentagon A and Pentagon B are congruent regular polygons. If the total length of the sides of Pentagon B is 68.5 feet, what is the length of each side of Pentagon A?

   13.7 feet

3. Is the following statement always true, sometimes true, or never true? Two congruent figures are similar figures. Explain.

   Always true; Possible answer: Congruent figures have the same shape, which is the definition of similar figures.

4. Draw a figure congruent to this line segment. Explain how you drew your congruent figure.

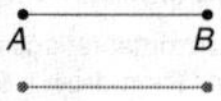

   I measured the line segment and drew mine the same length.

Circle the letter of the correct answer.

5. Which word makes this statement true? Corresponding parts of congruent figures are ______.
   - A not regular
   - (B) congruent
   - C polygons
   - D horizontal

6. If two angles of a right triangle are congruent, what are the measures of each angle in the triangle?
   - F 35°, 55°, and 90°
   - (G) 45°, 45°, and 90°
   - H 50°, 50°, and 90°
   - J 55°, 55°, and 90°

7. Which of the following polygons do not always have all congruent sides?
   - A a square
   - B an equilateral triangle
   - C a rhombus
   - (D) a pentagon

8. If ∠*A* of rectangle *ABCD* is congruent to ∠*X* of triangle *XYZ*, which of these statements is true?
   - F Rectangle *ABCD* is also a square.
   - (G) Triangle *XYZ* is a right triangle.
   - H Rectangle *ABCD* is a regular polygon.
   - J Triangle *XYZ* is an acute triangle.

LESSON 8-9

Reading Strategies

Graphic Organizer

This chart helps you understand **congruence**.

Definition	**Facts**
Two figures that have exactly the same size and the same shape are congruent.	• Figures have same size • Figures have same shape

Congruence

Examples

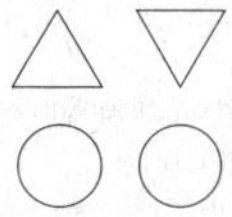

Non-examples

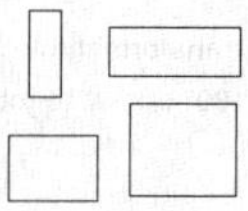

Use the chart to answer each question.

1. How do you know the triangles are congruent?

   They have the same size and shape.

2. Why aren't the rectangles congruent?

   They are not the same size.

Write *congruent* or *not congruent* to describe each pair of figures.

3. 

   not congruent

4. 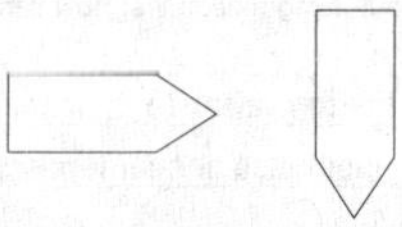

   congruent

5. 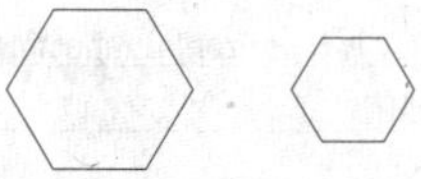

   not congruent

6. 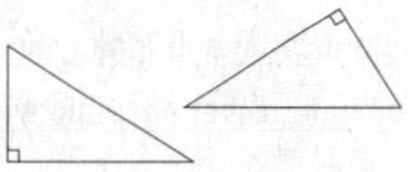

   congruent

LESSON 8-9

Puzzles, Twisters & Teasers

What is the difference?

Identify the pair of congruent figures in each row. Notice the letter next to each correct answer. Write the letter on the blank line which corresponds to the problem you have just solved. (There is only one correct answer per problem.)

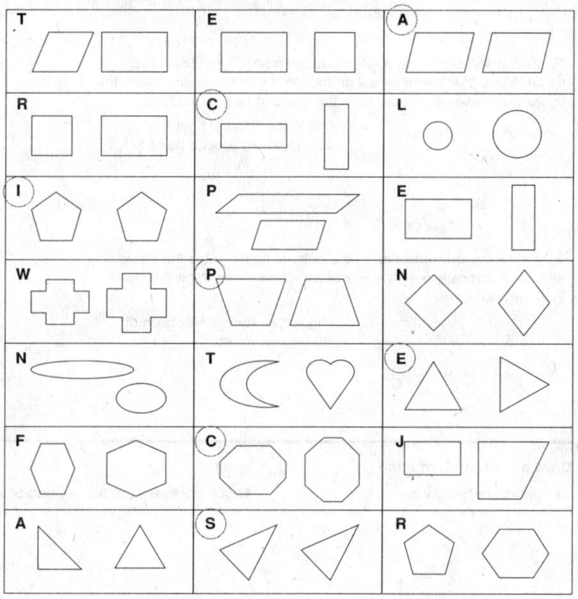

What do polar bears wear on their heads?

I C E   C A P S
3 2 5   6 1 4 7

LESSON 8-10

Practice A

Transformations

Tell whether each is a translation, rotation, or reflection.

1. 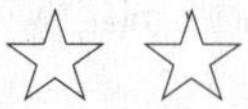

   translation, reflection, or rotation

2. 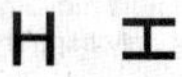

   rotation

3. 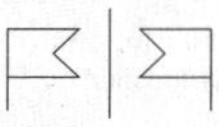

   reflection

4. 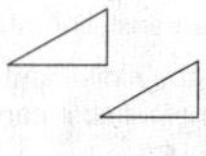

   translation

Circle the correct answer.

5. Which is the best description of the transformation shown below?

   - A a 90° clockwise rotation
   - B a vertical reflection
   - C a 180° clockwise rotation
   - (D) a horizontal reflection

6. Which is the best description of the transformation shown below?

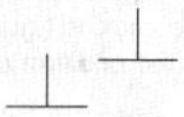

   - (F) a diagonal translation
   - G a vertical reflection
   - H a 45° rotation
   - J a horizontal reflection

7. How many degrees do you have to rotate any figure to get it back to its original position?

   360°

8. Why do words look different when you hold them up to a mirror?

   Possible answer: because the mirror shows a reflection of the words

LESSON **8-10**

Practice B
Transformations

Tell whether each is a translation, rotation, or reflection.

1.

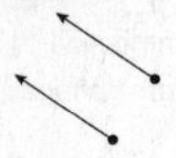

translation

2.

reflection

3.

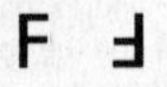

rotation

Draw each transformation.

4. Draw a 180° clockwise rotation about the point.

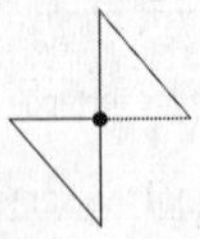

5. Draw a vertical reflection across the dotted line.

6. Without using reflections, how can you get this △ to look like this ▽?

Use the top vertex of △ as the point of rotation and rotate the triangle 180° clockwise or counterclockwise.

7. Describe a horizontal reflection of the word **MOM**. Can you think of any other words that would have a similar horizontal reflection? Possible answer:

The word MOM will look exactly the same if it is a horizontal reflection left or right. Other possible words: TOOT, HAH

LESSON **8-10**

Practice C
Transformations

Tell whether each of the transformations of each figure is a translation, rotation, or reflection.

1.

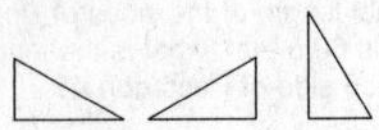

reflection; rotation

2.

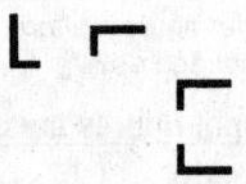

rotation; translation; reflection

Draw each transformation.

3. Draw a horizontal reflection across the dotted line. Then draw a 90° clockwise rotation of the reflected figure.

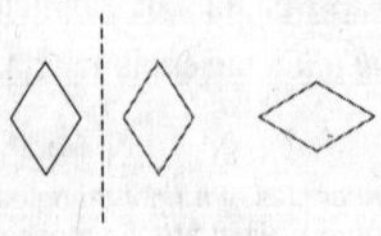

4. Draw a 180° counterclockwise rotation about the point. Then translate the rotated figure.

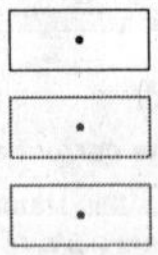

5. Name five capital letters that look the same before and after a reflection.

Possible answers: A, B, C, D, E, H, I, K, M, O, T, U, V, W, X, Y

6. Explain how folding a piece of paper could help you check your reflections.

Possible answer: I can fold the paper along the reflection line and hold it up to a light. The figures I see on both sides of the folded paper should match.

LESSON **8-10**

Reteach
Transformations

A transformation moves a figure without changing its size or shape. So, a figure and its transformation are always congruent.

A translation is the movement of a figure along a straight line. The location is the only thing that changes with a translation.

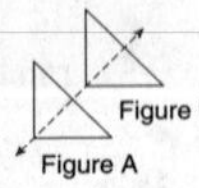

Figure B is a translation of Figure A.

A rotation is the movement of a figure around a point. The location and position of a figure can change with a rotation.

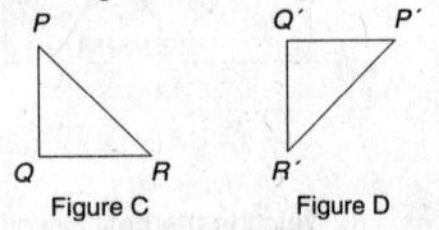

Figure D is a rotation of Figure C around point *R*.

When a figure flips over a line creating a mirror image, it is called a reflection. The line the figure flips over is called the line of reflection. The location and position of a figure change with a reflection.

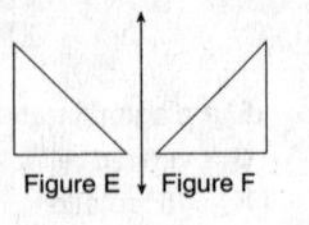

Figure F is a reflection of Figure E.

Identify each transformation.

1.

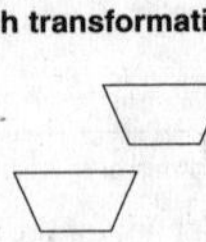

translation

2.

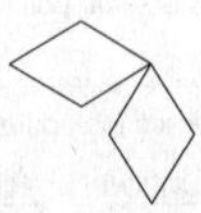

rotation or reflection

LESSON **8-10**

Reteach
Transformations (continued)

You can draw transformations.

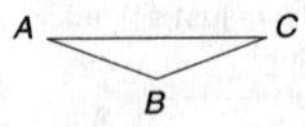

To draw a 45° clockwise rotation around point *C*, first trace the figure. Next, place your pencil on the point of rotation and rotate the figure clockwise 45°. Then trace the figure in its new location.

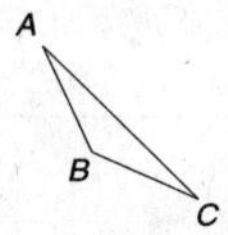

This is the original figure rotated 45° clockwise around point *C*.

To draw a vertical reflection of the original figure, trace the figure and flip it vertically over the line of reflection. Then trace the figure in its new location.

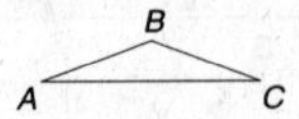

This is the vertical reflection of the original figure.

Draw each transformation.

3. vertical reflection

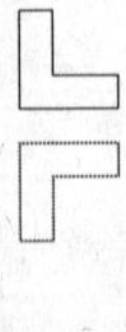

4. 60° clockwise rotation around point A

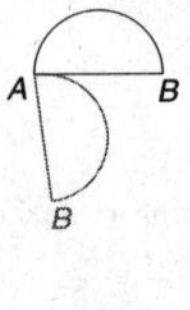

LESSON **8-10**

Challenge

Rotation Ride

The Ferris wheel shown below is turning clockwise around its rotation point at the center of the wheel. For each person on the Ferris wheel, describe how many degrees he or she must rotate to be able to get off the ride.

1. Darius 90°
2. Jenna 225°
3. Paul 0°
4. Sarah 315°
5. Mei 270°
6. Ben 135°
7. Mindy 45°
8. Carlos 180°

LESSON **8-10**

Problem Solving

Transformations

Write the correct answer.

1. If the rotation point of a circle is its center, how will all rotations affect the circle?

   The circle will never change.

2. What transformation could make an arrow pointing east become an arrow pointing north?

   a 90° counterclockwise rotation

3. What transformation could make the number 9 become the number 6?

   a vertical and a horizontal reflection or a 180° rotation

4. What transformation could make the letter P look like the letter b?

   a vertical reflection

5. On the coordinate plane at right, graph Triangle A with vertices (3, 1), (6, 1), and (3, 5). Then graph Triangle B with vertices (3, 6), (6, 6), and (3, 10). What transformation best describes the change from Triangle A to Triangle B?

   vertical translation

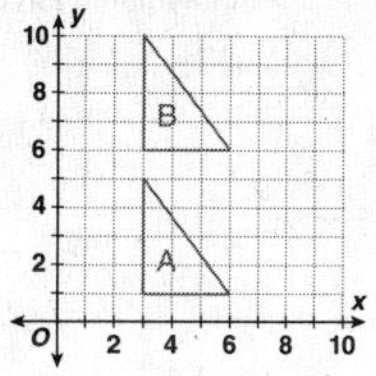

Circle the letter of the correct answer.

6. Which transformation best describes the figure on the right?

A 90° clockwise rotation
B horizontal reflection
(C) 90° counterclockwise rotation
D horizontal translation

7. Which transformation best describes the figure on the left?

(F) horizontal reflection
G 180° counterclockwise rotation
H 90° counterclockwise rotation
J horizontal translation

LESSON **8-10**

Reading Strategies

Focus on Vocabulary

Moving a figure does not change its size or its shape. **Transformation** means moving a figure. Here are three kinds of transformations.

- **Translate** → **Slide** a figure to a new position. Letter *P* has slid from one spot to a new spot.

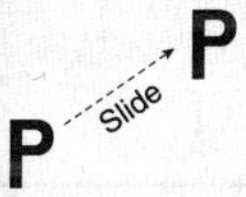

- **Rotate** → **Turn** a figure around a point to a new position. Letter *P* has been rotated to a new spot.

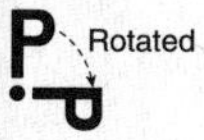

- **Reflect** → **Flip** a figure over a line to a new position. Letter *P* has been flipped over a line of reflection.

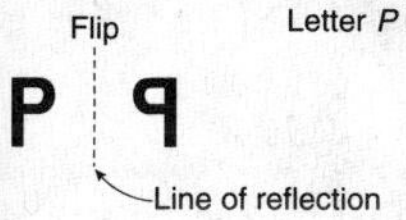

Write *translate, rotate,* or *reflect* to name the kind of transformation shown in each figure.

1. 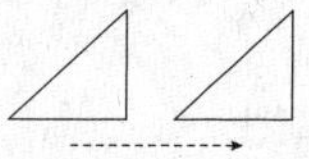

   translate

2. 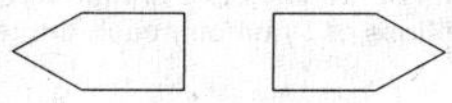

   reflect

LESSON **8-10**

Puzzles, Twisters & Teasers

Transformed

How would you complete each statement? Shade the correct box.

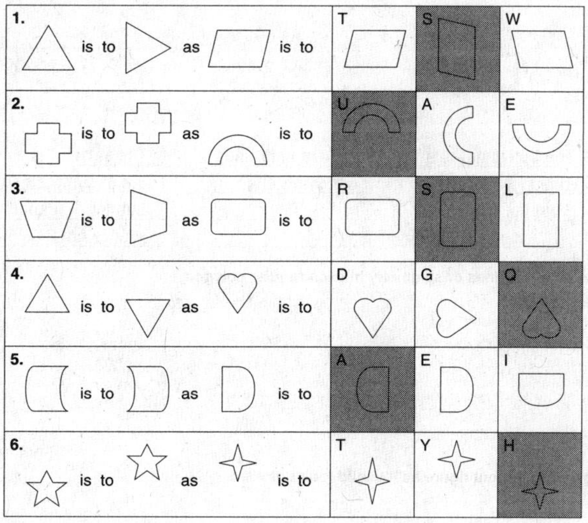

To solve the riddle, unscramble the letters in the shaded boxes.

What do you get when an elephant walks in your garden? squash

LESSON 8-11 Practice A
Line Symmetry

Determine whether each dashed line appears to be a line of symmetry.

1. 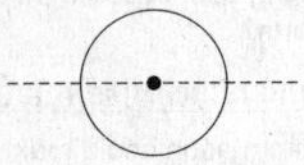

yes

2. 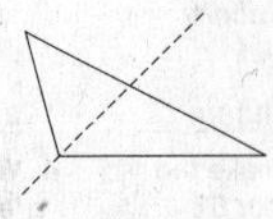

no

3. 

yes

Find all of the lines of symmetry in each regular polygon.

4. 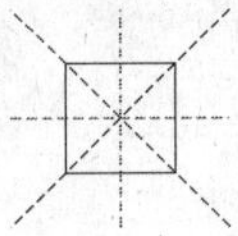

5. 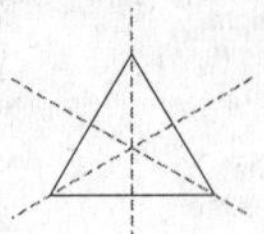

6. 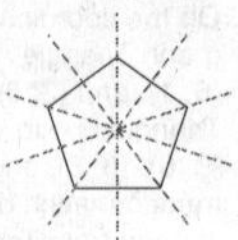

Draw each cut-out figure as it would look unfolded.

7. 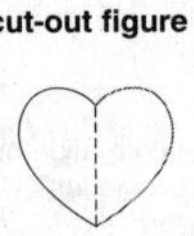

8. 

9. Draw a line of symmetry for the word.

10. How do you know if a figure has line symmetry?

Possible answer: If I can fold the figure and the two parts are congruent, then the figure has line symmetry.

LESSON 8-11 Practice B
Line Symmetry

Determine whether each dashed line appears to be a line of symmetry.

1. 

no

2. 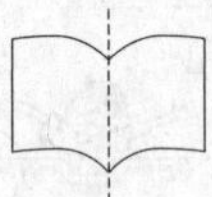

yes

3. 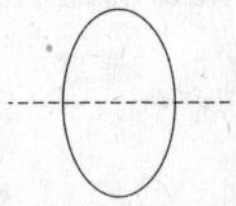

yes

Find all of the lines of symmetry in each regular polygon.

4. 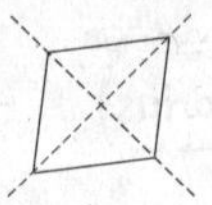

5. 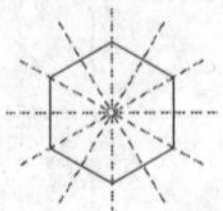

6. 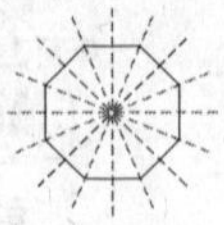

Draw each cut-out figure as it would look unfolded.

7. 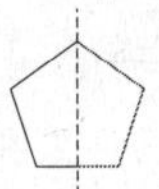

8. 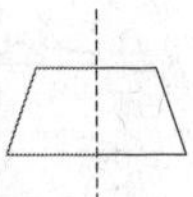

9. Which has more lines of symmetry, a square or a rectangle?

a square

10. Of the numbers 1 through 9, which numbers can have lines of symmetry?

3, and 8

LESSON 8-11 Practice C
Line Symmetry

Determine whether each dashed line appears to be a line of symmetry.

1. 

The horizontal is a line of symmetry, but the vertical is not.

2. 

They are both lines of symmetry.

3. 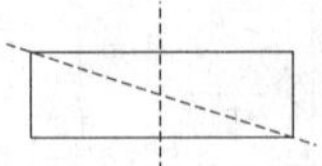

The vertical is a line of symmetry, but the diagonal is not.

Find all of the lines of symmetry in each regular polygon.

4. 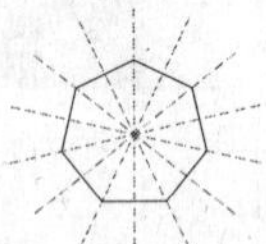

5. 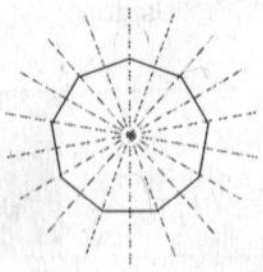

6. 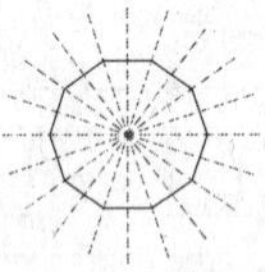

Draw each cut-out figure as it would look unfolded.

7. 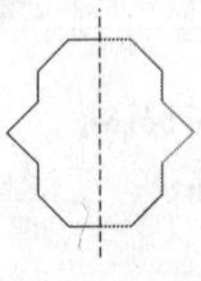

8. 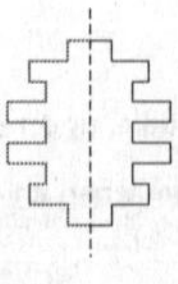

9. If all of the letters of a word have a horizontal line of symmetry through their centers, what would the word look like if it were vertically reflected?

It would look the same.

LESSON 8-11 Reteach
Line Symmetry

A figure has line symmetry if it can be folded or reflected so that the two overlapping parts of the figure are congruent. The line of reflection is called the line of symmetry.

To figure out if a dashed line is a line of symmetry, test to see if the two parts match exactly when folded or reflected across the line.

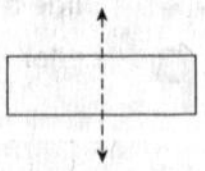

The two parts match, so the line is a line of symmetry.

Decide if each line is a line of symmetry.

1. 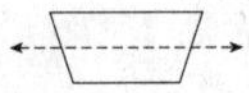

no

2. 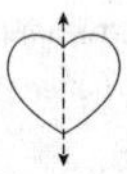

yes

To find all of the lines of symmetry in the regular octagon, first trace the polygon and cut it out.

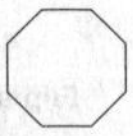

Then fold the polygon in half in different ways to find the lines of symmetry.

A regular octagon has 8 lines of symmetry.

Find all of the lines of symmetry in each figure. Write how many lines of symmetry each figure has.

3. 

1 line

4. 

4 lines

LESSON 8-11

Challenge

Alphabetical Symmetry

Tell whether each capital letter has a horizontal line of symmetry, a vertical line of symmetry, both, or neither. Draw the lines of symmetry to illustrate.

A vertical	B horizontal	C horizontal	D horizontal
E horizontal	F neither	G neither	H both
I both	J neither	K horizontal	L neither
M vertical	N neither	O both	P neither
Q neither	R neither	S neither	T vertical
U vertical	V vertical	W vertical	X both
Y vertical	Z neither		

LESSON 8-11

Problem Solving

Line Symmetry

Write the correct answer.

1. Do your body and face appear to have a vertical line of symmetry or a horizontal line of symmetry?

   vertical

2. Which letter of the alphabet has an infinite, or endless, number of lines of symmetry?

   the letter o

3. Ted says the diagonals of a rectangle are also its lines of symmetry. Do you agree? Explain.

   No, because when folded or reflected along a diagonal, the two parts of the rectangle do not match.

4. Using the digits 0 through 9 and not repeating any digits, write a 3-digit number that has a horizontal line of symmetry.

   3-digit numbers only choices are: 083, 038, 803, 830, 308, 380

5. Draw a line of symmetry for this word.

   DOCK

6. Draw the lines of symmetry for this star.

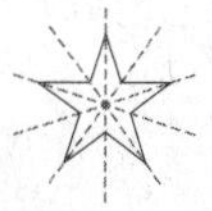

Circle the letter of the correct answer.

7. How many lines of symmetry does this hexagon have?

   A 4
   B 8
   ⓒ 6
   D 2

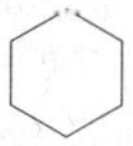

8. How many lines of symmetry does this flower have?

   F 3
   Ⓖ 4
   H 5
   J 6

9. How many lines of symmetry does a square have?

   A 0
   B 2
   ⓒ 4
   D 6

10. How many lines of symmetry does a regular pentagon have?

   F 1
   G 2
   H 4
   Ⓙ 5

LESSON 8-11

Reading Strategies

Focus On Examples and Non-Examples

If you fold a figure along a fold line and one side matches up exactly with the other side, the figure has **line symmetry**.

The fold line is called a **line of symmetry.**

Fold the rectangle along the diagonal line. The sides do not match up, so the fold line *is not* a line of symmetry.

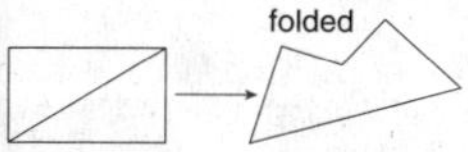

No Line Symmetry

Fold the rectangle along the vertical line. The sides match up exactly. So, the fold line *is* a line of symmetry.

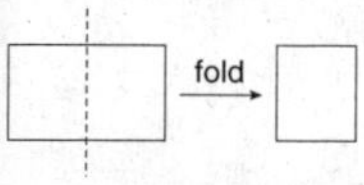

Line Symmetry

Trace and fold each figure below. If you can draw a line of symmetry, write *yes*. If not, write *no*.

1. 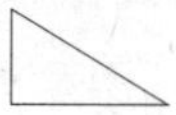

   no

2. 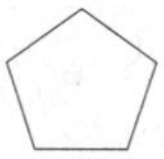

   yes

3. 

   yes

4. 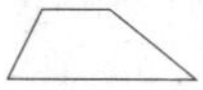

   no

LESSON 8-11

Puzzles, Twisters & Teasers

Mirror Images

Look at each letter. Circle the letters that have a line of symmetry, and then rearrange them to solve the riddle below.

N	(E)	(D)	P
(I)	(A)	G	R
(T)	L	S	(Y)
Q	F	(M)	Z

What is after sunrise and before sunset? DAYTIME